Rev. 07/22/2023

Feedback: alsnewideas@gmail.com

Contents

Preface

Albert Einstein and his second wife Elsa

This book seeks to provide lay readers the "Eureka!" moment of understanding Relativity without inducing headaches in comprehending the math. It is written for those who have studied it, wondered about it, and still lack an intuitive understanding of it. As well as science, it seeks to open a window into the intellectual spirit of Relativity that animated the minds of its discoverers, including Einstein, Poincare, Minkowski, and Lorentz.

Reading about Relativity as a non-professional physicist is frustrating. You feel it's an important aspect of the Universe you want to understand. But studying it either balks you with incomprehensible mathematical formulas or tries to placate you with platitudes of "everything is relative except the invariant speed of light" without explaining the how's and why's.

No matter how much information you take in, understanding remains elusive. You find yourself immersed in allegories such as Einstein's Train and Embankment, The Twins Paradox, and even The Andromeda Paradox, explained in contradictory ways. How can we glean an intuitive understanding of how time behaves in Relativity, when in everyday life we only experience time ticking at a constant rate and events pegged to a universally agreed upon timeline?

Einstein attempted to clarify Relativity in the public mind when he published ***The Special and General Theory of Relativity***, supposedly written for non-scientists, in 1916:

The present book is intended, as far as possible, to give an exact insight into the theory of Relativity to those readers who, from a general scientific and philosophical point of view, are interested in the theory, but who are not conversant with the mathematical apparatus of theoretical physics. – Albert Einstein

The English translation can be read on-line:

https://www.ibiblio.org/ebooks/Einstein/Einstein_Relativity.pdf

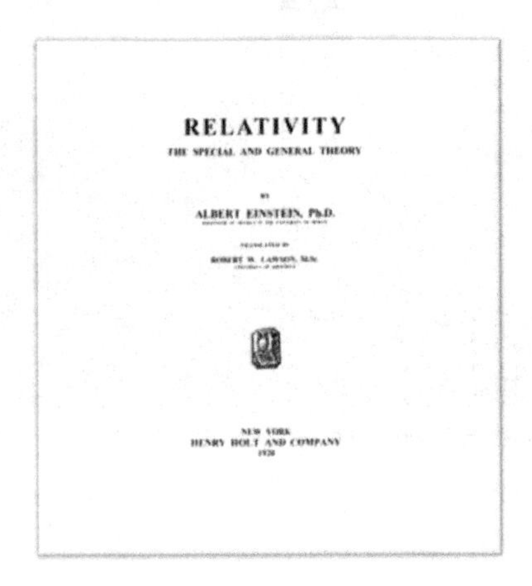

RELATIVITY

THE SPECIAL AND GENERAL THEORY

ALBERT EINSTEIN, Ph.D.

NEW YORK

HENRY HOLT AND COMPANY

Alas, try as he might, Einstein couldn't refrain from introducing "the mathematical apparatus:"

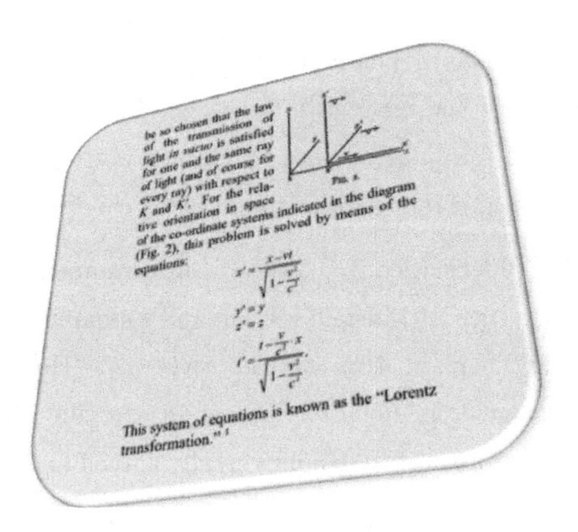

be so chosen that the law of the transmission of light *in vacuo* is satisfied for one and the same ray of light (and of course for every ray) with respect to K and K'. For the relative orientation in space of the co-ordinate systems indicated in the diagram (Fig. 2), this problem is solved by means of the equations:

$$x' = \frac{x - vt}{\sqrt{1 - \frac{v^2}{c^2}}}$$

$$y' = y$$

$$z' = z$$

$$t' = \frac{t - \frac{v}{c^2} \cdot x}{\sqrt{1 - \frac{v^2}{c^2}}}.$$

This system of equations is known as the "Lorentz transformation." [1]

Explaining Relativity this way is like describing an animal you've never seen by saying how much it weighs and how fast it runs. It obscures the fundamental understanding of the creature. I am explaining it as "Einstein's Car" in a way that keeps the discussion focused on common ways of thinking of Earthbound time and distance, while bringing Einstein's playful spirit to bear in understanding what lies beneath Relativity's veneer.

My study of Relativity began with the popular cosmology and physics books of the 1970s by authors like George Gamow and Isaac Asimov. My high school science project was the visual representation of Relativity's four dimensional Spacetime. With the advent of computers, I programmed three dimensional coordinates in engineering applications and four-dimensional Lorentzian coordinate systems in the study of Relativity. After a career in business centered around developing algorithms to control operations of multinational companies, I began writing on topics of history and economics.

I returned to the study of Relativity for four and a half years, wanting to visualize it as a holistic entity, rather than a mathematical abstraction. It would be pointless, and even insulting, to readers to merely rehash material available in popular physics books, the Internet, and Youtube videos, which after a lifetime of study I found remarkably unenlightening.

I restudied the theory from the original writings of Einstein and its other founders to the current state of information and belief. I progressed from the superficial familiarity gleaned from reading popular physics books to an intuitive understanding, the way one progresses in learning a new language --- from tedious thinking of how to translate it word by word, to intuitively hearing and speaking it without conscious translation.

My discussions with Relativity physicists lead me to believe my insights are worthwhile. I've been educated to aspects of Relativity I didn't understand, while helping physicists understand some so-called paradoxes by looking at Relativity in a broader way than its traditional dogma is taught. My reviews of recent popular physics books such as *Quantum Enigma*, *Why Does E = MC^2* and *How to Build a Time Machine* were well enough received to make me think it worthwhile to write a Relativity book that seeks to provide this understanding:

- *Wow! That review was awesome! I really appreciate it. I also intend to print it to have on hand as I read the book.*
- *I have already trudged through probably 20 other quantum physics books, and I kinda-sorta learned more from this review than many of my books.*
- *A great review! I've read perhaps two dozen books on am and still can't wait to read this one.... Loved your review.*

- *Really great summary- really helps to see the woods before possibly getting lost in the trees.*
- *Loved this review, the comments, and the book.*
- *What a stunning review! Thanks Alan Sewell. This is how all reviews should be.*
- *If you ever blog or podcast, I will totally buy a subscription.*
- *Alan, you have a gift to cut through the clutter and get straight to the point.*

While verifying that no one had previously used this title "Einstein's Car," I discovered a charming true story that only recently came to light. It shows Einstein touring in a mysterious car that fits the character of Relativity. I decided to explore the mysteries of Relativity by recreating Einstein's relativistic car.

Einstein's Car

In Las Vegas you can experience almost anything, from shooting machine guns to jumping off a 1,000-foot tower with a bungee cord. On your next trip there, you might see a billboard enticing you to "Drive the greatest cars of science!"

"Looks interesting," you say. "Let's see what that's all about."

"It's exactly what you think," says the proprietor. "You drive the cars of the world's greatest scientists."

"Weren't those guys pretty buttoned-down?"

"Some let their hair down," the proprietor assures you. "You'll see what I mean when you drive their cars. Our special today is Albert Einstein's Car."

You laugh. "Come on! Einstein never had a car. I'll bet he never even learned how to drive."

"Oh, you know his biography?" the proprietor says. "Then you'll enjoy the experience all the more. You're right about Einstein never owning a car. But he did drive a magnificent one. Let's take a look at it."

You smile a cynical grin and walk to the lot, where the proprietor points to the car he says Einstein drove. It's an antique luxury car in mint condition.

"That is a beautiful car!" you say.

"1931 Duesenberg," explains the proprietor. "A real collector's item. This one's worth nearly a million."

"And Einstein really drove it?"

“Here’s proof,” he says clicking on his smartphone “A Hollywood movie studio filmed it.” He clicks on the link to an on-line video:

You laugh again. “Well, OK, then, but I can’t drive a stick shift.”

“Don’t have to. We’ve added an autonomous driving program. It’s foolproof. Get in and explore the Universe the way Einstein saw it.”

“How much?” you ask.

He tells you.

“Wow! That’s a lot!”

“Take a spin and pay whatever you think it’s worth when you get back. I’m not worried, because most people think the experience is worth way more than I ask. Hop in, buckle up, and let the car do the rest.”

“You’re not going to bug me to pay if I don’t like it?”

“I’m a person of my word.”

He opens the door. You get in and buckle up. He enters a code on his smartphone to activate the engine. Lights start flashing on the console.

“Car leaves in two minutes,” says the proprietor. “Comes back to this spot when the tour’s over. The wheel’s disconnected because it’s self driving, but you can pretend you’re driving, if you want.”

He closes the door, waves you off, and walks away. You shake your head and laugh at how you let yourself be hornswoggled, like any dumb tourist. But he promised you wouldn’t have to pay if you didn’t like it, so what the hey? You’re mesmerized by the flashing lights rolling across the console and don’t notice the two minutes have ticked down.

Sure enough, the car wheels out of the parking lot and goes careening down Las Vegas Boulevard. The experience is spooky from the gitgo. Every car coming up behind you, coming toward you, and crossing from the side appears to be closing distance with you at the same speed. No matter how much Einstein's Car twists and turns, every other car on the road overtakes you then pulls away with speeding up or slowing down. Even when Einstein's Car stops at red lights, the other cars careen around you, front, back, and sideways, at constant speed, somehow never hitting you.

Einstein's Car drives onto the expressway. Other cars keep coming from behind and in front of you at the same speed as if you're standing still, while you're passing landmarks on the ground at increasing speed. As you approach 60 mph, the horizon ahead takes on a futuristic hue. You see oscillating bands of light and darkness scudding across the land as the days and nights of the future cast their sunbeams and shadows. New constructions sprout from the land and unfamiliar aerial craft crowd the skies ahead. You swivel around to see what's going down behind you. You see buildings de-constructing into simpler forms, then disappearing. The farther back you look, the simpler the constructions are. On the far backward horizon you see nothing but desert and cactus. Then a bit of movement. Was that an Indian riding in war paint on a horse?

Einstein's Car drives itself back to the lot. The door opens and the proprietor is there waiting for you.

"How'd you like it?"

"Quite a show," you answer. "You had me buffaloed, until I realized it was a simulation. You put movie screens around the car when I was looking at the flashing lights inside, then you projected the show. Clever trick, I'll grant you that."

The proprietor frowns. "I didn't put any movie screens around the car. You saw the world --- the real world --- the way Einstein saw it. Now comes the educational part of the program. You saw other cars passing you at the same speed, no matter what speed you were going, or in which direction. When you looked in front, you peeked into the future. When you looked backward, you saw the past. If you can explain it, the way Einstein would have, I'll give you a framed copy of Einstein's 1921 Nobel Prize in Physics to take home."

"Time dilation!" you say, beaming. "Everyone knows Einstein explained how time slows down the faster you go. It works the same when I'm talking to my accountant. A minute with her seems like an hour. It's billed for an hour, too."

"Ha!" says the proprietor. "You got some of it right. Time dilation is important, but in Relativity multiple things are happening all at once. That's what makes it so confusing. Time dilation isn't the reason the other cars always passed you at the same speed, or why you could peer into the past and future."

"You seem to know a lot about it."

"I've done my homework with all these cars. I have another one driven by Erwin Schrodinger, the quantum mechanics wizard. Drive that one, and you won't know where it's going 'til it gets there."

"So, what's wrong with my answer about time dilation?"

"Time dilation is trigonometric. It doesn't really kick in until you're going 80% the speed of light or better. The other cars passed you at constant speed at whatever speed you were going, even when you were stopped. There had to be an effect erasing speed differentials between Einstein's Car and every other car on the road, regardless of how fast you were going, or in what direction. That's not time dilation."

"Hmmmm.... I'll have to think on that for a while."

"If you come up with the correct answer before closing time tomorrow, I'll give you the prize. My deal is what I said it was. Pay me what I asked before you get in the car or pay me what you think it's worth."

You walk with him inside, so distracted that you pay the asking price for the ride without bothering to haggle. "At least it's the classy way to fleece people," you think, while still wondering what makes Einstein's Car tick. Einstein's Car, however it operated, has set you to thinking about something fundamental you'd always wanted to understand but never quite grasped. Now you can't let go of it. You think about going back to your hotel and doing some internet research to see if you can discover the mystical effect the proprietor was talking about. Having a copy of Einstein's Nobel Price would be a great conversation piece.

After selecting the concept and presumptive title for this book, I checked the Internet to make sure nobody else had thought of it. When I searched on "Einstein's Car" I discovered that Einstein really took a mystical ride in a 1931 car. The car took a relativistic trip around the world in a few minutes, even soaring through the air. He "drove" it at Universal Studios in California. It was discovered hidden away in the dusty shelves of Universal Studio's warehouse of archived movies in 2018. The movie showcases Einstein's lighthearted character that made him an endearing personality as well as the preeminent theoretical physicists of his century, and maybe of all time:

https://www.youtube.com/watch?v=bKuR2BnnQy4

Let's explore his view of what makes the Universe tick, as seen from his fantastic car.

The Relativity of Simultaneity

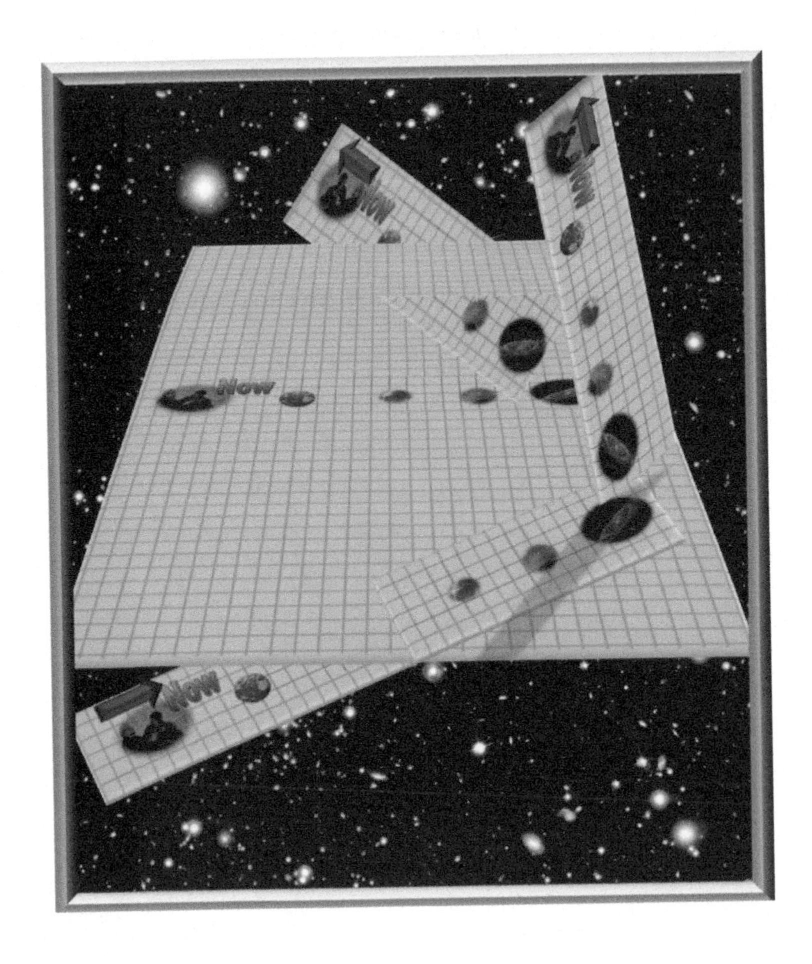

In the 1880s, physicists A.A. Michelson and Edward Morley designed optical devices to measure the speed of light. They expected to measure directional differences resulting from the Earth's motion through space. However, they found the velocity of light constant in all directions, regardless of motion of the sender and receiver. You can't gain or lose speed on a flash of light no matter the speed or direction you twist and turn between the time it is emitted and observed.

Einstein expressed the physicists' astonishment at discovering the speed of light has no directional component to its velocity:

"If the Michelson–Morley experiment had not brought us into serious embarrassment, no one would have regarded the relativity theory as a (halfway) redemption."

This is what Relativity ***is:*** the explanation of how the speed of light and all other fundamental forces can remain constant for everything in the Universe, regardless of each object's motion relative to every other object. It would be like every terrestrial object measuring a constant windspeed regardless of how fast or slow it travels or in which direction. Or a boat always finding the waves approaching or receding at the same speed, regardless of how the boat is moving; and that every boat on the water --- fast, slow, or stationary --- measures the waves as having identical speed.

Why is the speed of light and other fundamental forces constant for all observers, regardless of their motion and direction? Perhaps because of the fundamental mass to energy equation $e = mc^2$ -- also written as $m = e/c^2$. If the speed of light and other forces varied according to relative motion between the emitting and receiving objects, c would no longer be constant, and neither would m. Each object's mass would vary according to your motion toward or away from it.

Because gravity is proportional to mass, the gravitational force would vary with the changing relative motions of objects, destabilizing the orbits of everything bound by gravity. The imbalance would become extreme within atoms, where nuclear forces would vary with each particle's changing relative motion. Matter is energy bound by nuclear forces. If nuclear forces became unbalanced by motion, atoms would fly apart and matter could not exist. Relativity was first developed to explain the electrodynamics of atomic forces emitted by particles moving near light speed. It acquired a cosmological character later, after the puzzling results of the Michelson-Morley experiment were known. Having a constant speed of light and other forces regardless of motions of the emitting and receiving objects is therefore necessary for the Universe to exist.

Now the question becomes ***how*** can light and other forces maintain constant speed between the transmitting and receiving objects, no matter how much they bob and weave with different speeds and directions while the photons and force-carrying particles are in transit between them?

The answer often recited is: "The speed of light is measured to be constant because a moving measuring device shrinks if it is a mechanical device due to length contraction or is slowed by time dilation if it is electronic. The adjustment of the measuring device makes the speed of light constant for the measurer."

But that's not it. Time dilation and length contraction derive from the trigonometric function designated "Gamma" that doesn't become significant until an object is moving near the speed of light. At lower speeds --- for example, an object observed moving at 10% the speed of light --- Gamma is only about .5% (one half of one percent), so there must be something else that adjusts the other 9.5% of the speed differential between the moving object and light. At ordinary speeds, Gamma is essentially zero, but the speed of light still has to be adjusted to make it constant when objects are moving relative to each other, even if it is only somebody walking past you.

In looking at how the Universe adjusts itself to assure the speed of light is constant for all observers in all circumstances of motion, let's note that speed is distance divided by time. If the distance between the sender and receiver changes while a photon of light or a particle of force is in transit, then to maintain a constant speed, the time of transmission of the photons and other force-carrying particles must change to offset the relative motion between sender and receiver. As Albert Einstein told us:

Every reference body (co-ordinate system) has its own particular time; unless we are told the reference-body to which the statement of time refers, there is no meaning in a statement of the time of an event.

He meant that each object moves in its own velocity frame, having a unique "it's happening ***now***" moment in other objects' past or future. The chapter heading picture shows various incarnations of Albert and Elsa experiencing different "now" moments. Objects only have the same "now" if they are in contact with each other at the same point in space, or if separated, are aligned with identical velocities and directions. Even people walking past each other on Earth have "now's" differing by nanoseconds (billionths of a second). The speed of light is kept constant for all observers by making their "now's" ***inconstant.***

Relativity of Simultaneity is the mechanism for bending time backward and forward as the means for propagating light and other forces through space at constant velocity, regardless of motions of the sender and receiver. This most fundamental principle of Relativity is often glossed over in popular literature, in favor of discussing time dilation. Perhaps this is because "Relativity of Simultaneity sounds pedestrian. It may have garnered more interest if properly called ***The Relativity of Events.***

An ***event*** is any action that emits photons or force-carrying particles that interact with other objects in the Universe, providing information that the event took place at a particular place and time. Events emitting photons and force-carrying particles may trigger events in the objects receiving them --- such as photons emitted from a fire igniting other fires at a distance --- creating expanding causality chains spreading outward at the speed of light. Anything affecting the "now" moment of the original event will affect the "now" moments of events further down the causality chain. Time travel mavens hope to put Relativity of Simultaneity in harness to get into an event's past and change or inhibit its effect on the Universe.

I will explain Relativity of Simultaneity using Einstein's Car as a proxy for objects absorbing photons or other force-carrying particles. I will use Yellow Car as a proxy for photons and force-carrying particles travelling at constant speed of 60mph, a car-centric analog for the speed of light. We can anticipate what might happen at faster-than-light speeds when Einstein uses his car phone (a radiophone in his day) to communicate at what amounts to infinite speed compared to our 60mph car-centric speed of light.

Relativity of Simultaneity is one of two components of the Lorentz Transformation mathematically describing how objects in motion measure the same constant speed of light originating from other objects in various states of motion. The other component is the Gamma Function describing time dilation. Because Relativity of Simultaneity is the dominant component of the Lorentz Transformation in velocities below about 85% of light speed, I will explain it first, by setting Einstein's Car moving at 30mph, half the 60mph car-centric analog of lightspeed. The Gamma function at this speed would multiply Simultaneity by about 1.1547. I will not consider that for now, because there is more clarity in studying Relativity of Simultaneity in isolation, then combining it with Gamma in the chapter on The Lorentz Transformation.

Relativistic vs. Newtonian Motion

The simplest case of motion, Newtonian and Relativistic, is when Einstein's car is stationary relative to the garages. Let's say Yellow Cars leave both garages simultaneously and move toward Einstein's Car parked eight miles away at a constant velocity of 60mph. Both Yellow Cars will travel the eight miles to Einstein's car in eight minutes:

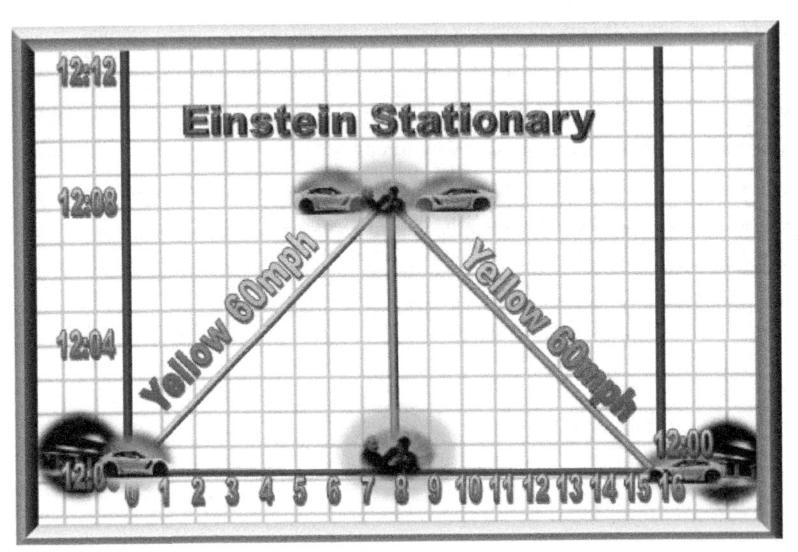

Now let's look at how Yellow Cars leave their garages at 12:00 and move in converging directions at 60mph while Einstein is moving left-to-right at 30mph. This sequence of Newtonian overtake and collision is:

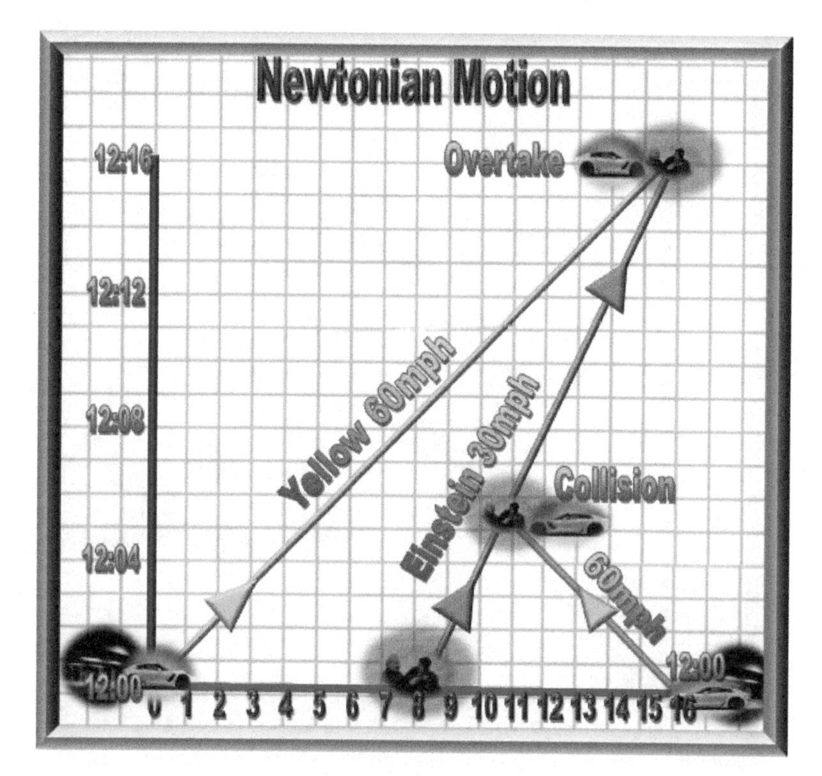

Yellow Car overtaking Einstein's Car from the left has a speed differential of 60pmh – 30mph = 30mph. The initial 8-mile separation gives Einstein's car enough head start to get eight more miles down the road before Yellow Car overtakes it at 12:16 at mile 16.

For collision (we hope a near miss) of Yellow Car approaching from the right, the combined closing speed is 60mph + 30mph = 90 mph, or 1.5 miles per minute. Because Yellow Car is moving twice as fast as Einstein's Car, the point of collision computed from Einstein's direction of travel will be 8 miles + 1/3rd * 8 miles = mile 10.67. The same as computed from Yellow Car's direction of travel as 16 miles – 2/3rd * 8 miles = 10.67 miles. The collision will take place at 8 minutes * 2/3rd = 5.33 minutes or 12:05:20.

Relativistic overtake and collision profoundly differ from their Newtonian counterparts because there cannot be a speed differential of 60mph – 30mph on the overtake or 60mph + 30mph on the collision:

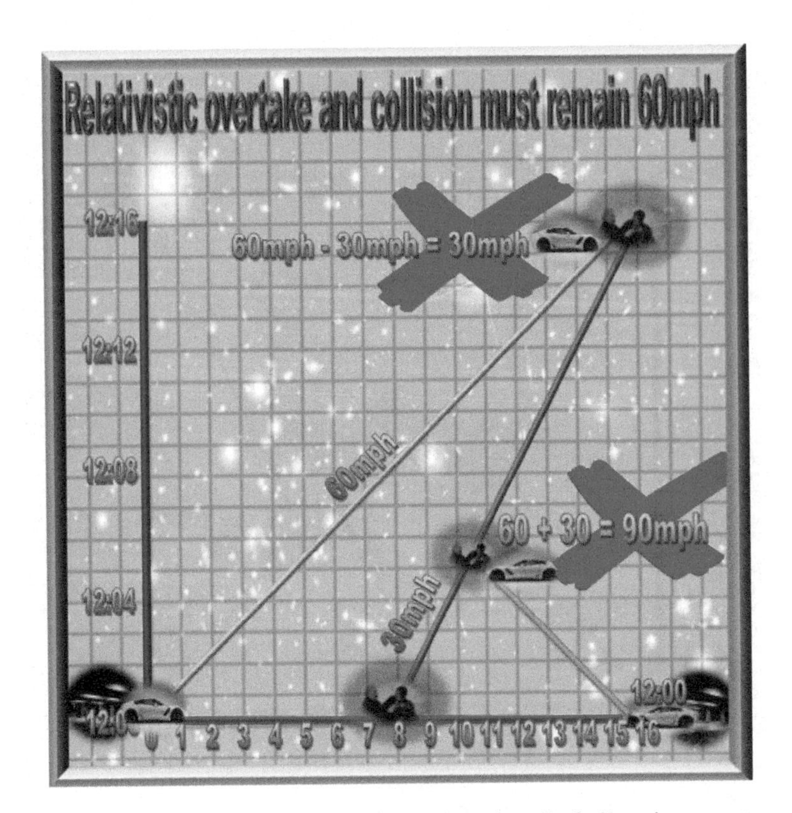

Because the speed of light is constant, Yellow Cars from both directions must reach Einstein at constant 60mph, as our car-centric speed of light. To understand this Relativistic motion through time and space, it is necessary to consider each object stationary as the Universe moves around it, while the time of origin from the sending object advances or regresses while the photon or force-carrying particle (Yellow Car in our analogy) is in flight. We will therefore align our perspective to make Einstein' car stationary, as it would be if we move with him and Elsa:

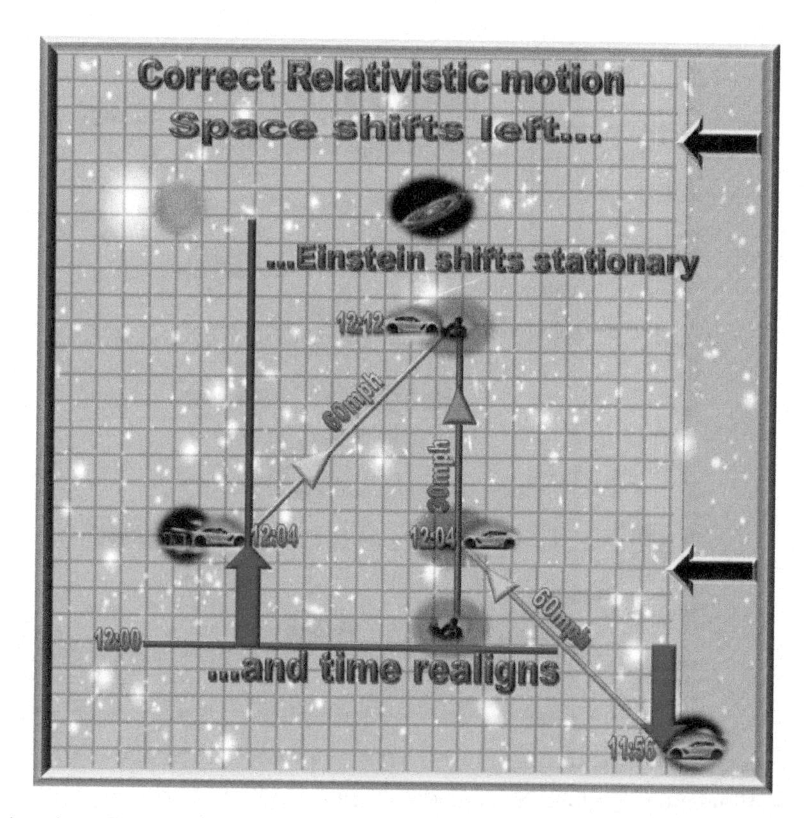

We've given Einstein a stationary perspective by shifting the graph to the left as he travels. To keep this speed constant while Einstein moves, Relativity of Simultaneity has to shift Yellow Cars' ***times*** of leaving the garage.

Yellow Car Left will leave the garage four minutes later at 12:04 (as seen by Einstein) while travelling to overtake the Einsteins. For Yellow Car Right, the time of leaving the garage regresses four minutes earlier at 11:56, as seen by the Einsteins. The constant speed of light requires that if Yellow Car Left and Yellow Car Right took eight minutes at 60mph to get to Einstein while he was stationary relative to them, they must take eight minutes at 60mph to get to him any way he moves toward or away from them while the Yellow Cars are in flight.

At the end of the overtake and collision sequence, Einstein might conclude, "My car and the Yellow Cars left at 12:00. By the time Yellow Car from the left overtook me, it left four minutes later than me at 12:04. By the time Yellow Car from the right passed near me, it left four minutes earlier at 11:56." When Relativity of Simultaneity adjusts the times of origin this way, the compounding Newtonian overtake curve goes away:

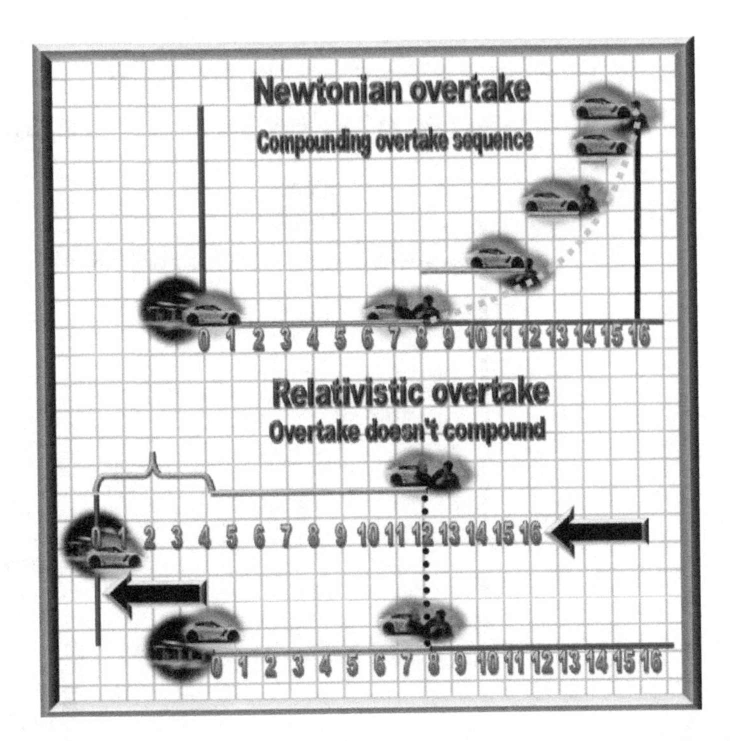

In Newtonian overtake, there is a compounding sequence of 8 + 4 + 2 + 1 + .5 + .25 + .125 + ... minutes = 16 minutes. By the time Yellow Car gets to mile 8, Einstein's Car has moved on another 4 miles travelling at half of Yellow's speed. Then Yellow Car must traverse those 4 miles, giving Einstein's car another 2 miles to travel ahead, requiring another mile of overtake, and so on. In Newtonian overtake, if Einstein's Car is travelling at 59.999mph per hour, it will take 8,000 hours (almost a year) for Yellow Car going 60mph to overtake it at 480,000 miles.

In Relativistic overtake, it is just the original separation distance plus the percentage of light speed. If Einstein is travelling at half the car-centric speed of light, the overtake is 60mph for 8 miles plus 4 miles of simultaneity = 12 miles in 12 minutes at constant 60 mph. If Einstein's Car were travelling 59.999mh in Relativistic overtake, it will be overtaken by Yellow Car in 15.999miles in 15.999 minutes. Relativistic overtake cannot exceed 2x the distance of original separation.

A similar adjustment is made for the case of collision, whereby the time of origin of Yellow Car leaving the garage is regressed four minutes to reduce the closing speed from a combined 60 + 30 to just 60 mph.

Let's now combine the two Einsteins, Green Einstein stationary relative to the garages, and Red Einstein moving away from the one on the left and toward the one on the right at 30mph. Let's show it in two parts: the top is the way stationary Green Einstein sees his "now" moments timeline and the bottom the way moving Red Einstein sees his:

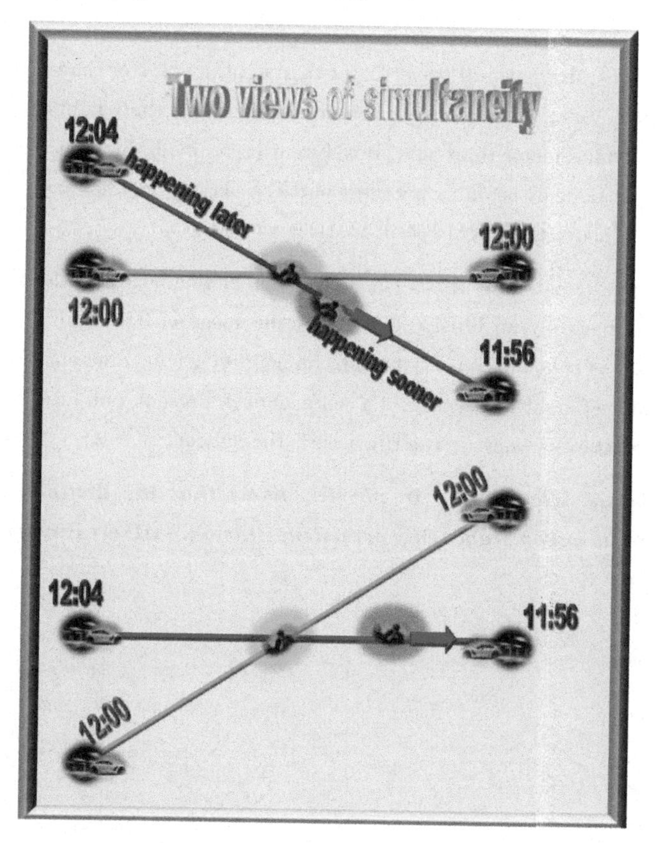

The different timelines of Yellow Cars leaving the garages opens a window on the possibility of time travel, or at least communications, between timelines. If the Red and Green Einsteins have an instantaneous communications device, like a radiophone, to talk to themselves and the garage attendants, then Red Einstein could can talk to Green Einstein and learn that Yellow Car Left will leave four minutes earlier than on Red Einstein's timeline. Red Einstein will know Yellow Car has left the garage while it is still parked on Red's timeline. Green Einstein could likewise learn from Red Einstein that Yellow Car Right has left the garage four minutes before it will leave on Green's timeline. Red and Green Einstein could harness their relative motion to get a peak into each other's future events ***if*** they had an instantaneous communications device.

Some aspects of hypothetical time travel by Relativity of Simultaneity are:

1) As a linear function, objects don't need to move anywhere near the speed of light to diverge their timelines. Thus, no need to expend those prodigious energies required to move objects to near light speed.
2) It doesn't require physical travel faster than light, just faster than light communications. Of course, we're no closer to communicating faster than light than we are moving physical bodies faster than light. But if we ever do break the light barrier, it seems more likely to be done by beaming massless particles across timelines than accelerating objects of mass to faster-than-light speed.

However, there remains the fundamental question of whether the Relativity of Simultaneity is intrinsically real or merely an illusion created by the speed of light being the mechanism that forces the *appearance* of constant velocity back to its point of origin. Einstein believed simultaneity planes are real and our Earthly perception of a common past, present, and future experienced as one timeline by every object everywhere in the Universe is the illusion:

People like us, who believe in physics, know that the distinction between past, present, and future is only a stubbornly persistent illusion. – Albert Einstein

Propagation vs. Simultaneity

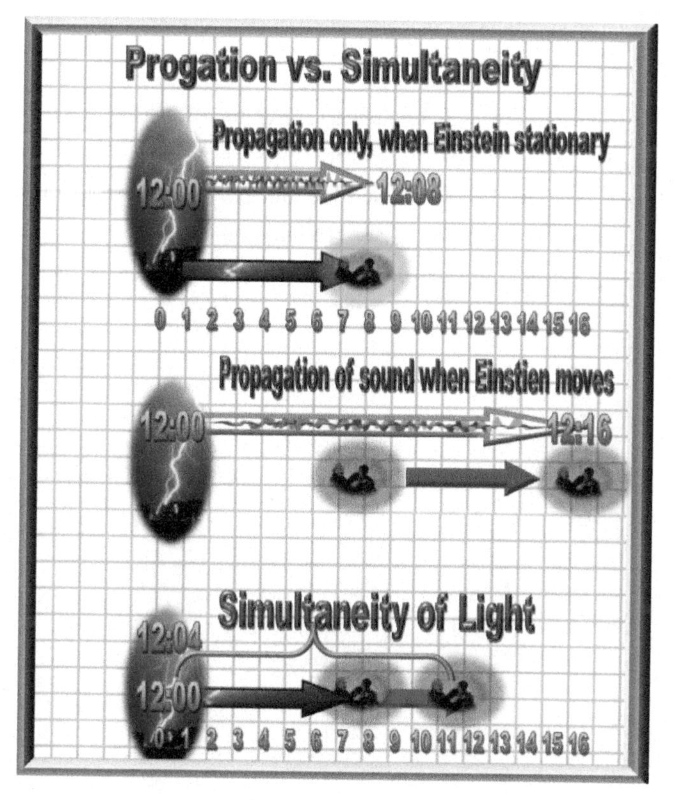

To determine whether Relativity of Simultaneity is intrinsically real or merely a subjective perspective --- such as arbitrarily setting your clock forward six hours and claiming you're seeing a "midnight sun" --- we need to know whether it is a propagation phenomenon, such as a thunder clap you hear later if you're moving away from a lightning strike, or whether it's something more fundamental that changes the nature of time. The chapter heading diagram compares the different propagation and simultaneity properties of light and sound by assuming they both move at 60mph.

The topmost frame shows Einstein stationary 8 miles away from a lighting strike. Light and sound, both moving at 60mph in our car-centric world, reach him 8 minutes later.

In the middle frame, Einstein starts moving away at 30mph as soon as the lighting bolt hits. The sound of the thunder, being a propagation phenomenon, overtakes him in the Newtonian way at 12:16 and mile 16.

In the lower frame, light overtakes Einstein in the Relativistic way as a propagation + simultaneity phenomenon, whereby the time of the strike advances from 12:00 to 12:04 as Einstein pulls away from it, overtaking him at 12:12 at mile 12. The propagation part is 8 miles and 8 minutes, the same as it would take to reach Green Einstein at rest. The simultaneity difference is the 4 minutes and 4 miles after that.

Relativity of Simultaneity of light appears to change the time an event occurs, whereas Relativity of Propagation of sound just changes the time you hear the noise, without affecting the time of the event that made it. There is something about light that wraps time around it like a mantle and carries it along. Because Relativity of Simultaneity appears to change the times when events happen, it might become our channel into moving information between past, present, and future.

Future Falling, Past Rising

Relativity of Simultaneity theorizes that objects moving toward an event experience the event happening sooner than objects stationary to it. Objects moving away from an event experience it happening later. In the chapter heading diagram, Einstein is speeding through contemporary Chicago. His velocity pulls the future incarnations of the city down into his present moment, while pulling the past incarnations up into his present. If he had an instantaneous communications device, he'd be able to pass information along the red line from Chicagoans of the future to Chicagoans of the past, assuming they also had instantaneous communications devices to talk to him.

In real life, we never see events in the future and past happening in our "now" moment because we don't have communications devices that send and receive information faster than light.

By the time the light from a future event reaches us, the event that spawned it has already sunk down into the past. If we were travelling at half the real speed of light, we might theoretically be able to see 4 minutes into the future of the Sun at 8 light minutes away; 2.15 years into the future of Alpha Centaury 4.3 light years away, and 1.25 million light years into the future of the Andromeda Galaxy 2.5 million light years away. But since it takes twice the 4 future minutes, twice the 2.15 years, and twice the 1.25 million years for light to reach us from those points, the event has already happened in the timelines of those places and sunk down into their past before we know anything about it. By the time we could get word to them about a future event in their timeline, it has become ancient history.

To see the future before it happens or preempt a past event that's already happened, we'd have to communicate at least twice the speed of light. We'd have to receive information about an event in the future, then send a message to the future people, informing them it will happen before it does. Same process if we wanted to send a message from our present into the past to persuade someone there to preempt an event we know will be detrimental. Unless we could send the message more than twice as fast as light, the past event we want to change or cancel will fall further pastward faster than our signal can reach it. Ideally, we'd want to communicate instantaneously with the future and past so we wouldn't have to worry through propagation delays.

What might happen if we *could* send and receive information instantaneously? Let's imagine how that might pan out by introducing communications by radiotelephone into the Einstein's Car scenario. Let's imagine three sets of Einsteins in radiotelephone communications with the city of Indianapolis. Green Einstein is stationary; Red Einstein is driving toward Indy, while Yellow Einstein is driving away. All are in radiotelephone contact with Indy and with each other. The following conversation takes place:

Einstein Green receives a message from the weather station in Indy that a big storm is brewing up. The weather alert staff want to know if it poses a danger to the public. Einstein Green calls Einstein Red whose Simultaneity moment lies a couple days in the future.

"A storm is brewing up over Indy," Einstein Green tells Einstein Red. "They want to know if they need to take precautions."

Einstein Red calls the Indianapolis weather station in the future and is told an occluded front has stalled over the city, raining buckets of water, and causing a highly destructive flood with widespread property damage and loss of life. Einstein Red reports the future information to Einstein Green, who relays it to Einstein Yellow whose timeline is to the city's past, where the weather station is reporting sunny blue skies. Einstein Yellow tells the weather alert staff that in a couple days the city will be underwater, so they need to get busy sandbagging the properties near the river and evacuate people living in low-lying areas. A few days later, the flood hits hard, but people have been warned. Properties have been sandbagged, moveable items of value relocated, and people in peril evacuated, so no lives were lost. A potentially deadly future event has been tamed. What else might we do with this power, not only over nature, but of future nature?

Macrocosm #1: The Cosmic Time Loop

Perhaps we might harness Relativity of Simultaneity in a big way by extending it across the Cosmos, if only we had instantaneous communications faster than light. The time loop conjectured above has a mother ship launching a fleet of satellite ships toward Earth, the satellite ships connected to each other with instantaneous faster-than-light communications, with one always passing near Earth to communicate with Earthlings of the present. The "ships" don't have to be large. They could be transceiver microchips the size of postage stamps accelerated to high velocity by beamed energy from the mother ship. As they approach Earth, they connect with Earths-of-the-future. As they speed past Earth, they connect with Earths-of-the-past. Whatever ship is closest to Earth at any moment would be communicating to Earthlings of the present time, because even though it is moving fast, it is close to Earth, so the Relativity of Simultaneity difference is not leveraged beyond the present moment. A transceiver could be set up on Earth-Now that receives information coming in from Earths-past and Earths-future and retransmits it in both directions. Maybe we could get communications going between our ancestors, ourselves, and our future descendants.

Earthlings-past wouldn't have communications devices to receive the signals relayed from Earthlings-future, but maybe we could communicate with them by focusing a powerful FTL beam on Earths-past, lighting them up in a sort of Morse Code. Earthlings-past would see lights in the sky flickering in patterns and learn to decode them. Could that be what we're seeing with those mysterious UFO's said to be sashaying across the skies? Earthlings from early hominid times until the far future might whoop it up together, changing Humankind's history for the better --- if only we had an FTL communications device to harness the simultaneity planes.

Because Relativity of Simultaneity theoretically links the past, present, and future into an integrated whole, there is a "Block-time" theory surmising the Universe was created fully grown from the Big Bang, such that everything that has happened, is happening now, and will happen in the future simultaneously coexist:

The block-time Universe theory is touted in some philosophical and psychological circles to deny that human beings have free will to make conscious decisions. If the past, present, and future are predetermined, how can we make decisions to change a future that already exists? This theory appeals to people who want to excuse themselves and others from the consequences of their bad decisions. They claim that if the Big Bang determined 14 billion years ago that we were going to do something bad now, how can we be held responsible for doing it? However, Stephen Hawking, the preeminent cosmologist of our time, was skeptical: "Do you notice how people who claim to believe in a pre-determined future still look both ways before crossing the street?"

Macrocosm #2: The Andromeda Paradox

Physicist and mathematician Sir Roger Penrose conceived The Andromeda Paradox to imagine the fantastic leverage Relativity of Simultaneity might have over extreme distances. In this paradox, an Earthling (let's call him Ernie) communicates with the past and present incarnations of an alien (let's call her Andrea) in the Andromeda Galaxy by walking in different directions. This scenario requires a Cosmic Phone that communicates between Earth and Andrea's planet in Andromeda instantaneously, without the tedious 2.5 million propagation delay in each direction required by plain-vanilla light. Assuming such fantastic communications could be created, the paradox goes like this:

Earthman Ernie gets a call on his Cosmic Phone from his trans-galactic girlfriend Andrea in Andromeda.

"Wha's up, Andie?"

"Nothing good," she responds. "Our Galactic Council has convened another emergency session to discuss invading Earth."

"What did we do this time?" asks Ernie.

"I don't know.... maybe they're just bored and looking for new worlds to conquer. Can you peek ahead and see what they decide, then I'll see what I can do to head 'em off, if I have to."

"Yeah, let me see what's goin' down in five days," sighs Ernie. "If it's as bad as we think, I'll past-track the information to you, so you'll have time to work on them."

He gets up and walks toward the direction he knows Andromeda is in this time of day. His simultaneity plane with Andrea-now, leveraged 2.5 million light years, moves forward five days into Andrea's future. "Still there?" he asks.

"Well, they did it," Andrea Future answers. "Our ships are on the way to attack Earth."

"OK," intones Ernie grimly. "Let me past-track the information and see if you can stop 'em."

He turns around and walks in the other direction, away from Andromeda. His plane of simultaneity reverses five days, past Andrea Now's timeline, and goes down another five days into Andrea Past's.

"Trans-Galactic greetings, dearest one," says Ernie. "I need your help on something."

"What is it?" asks Andrea Past, alarmed by Ernie's tone.

"I got a message you'll be sending me five days from now. At that time, you're going to advise me that your government has convened to decide whether to attack Earth. I talked to *Future You* five days after that, and you said your government decided to go through with it. Their ships are on the way to attack Earth right now."

"Sounds like something they'd do," acknowledges Andrea Past. "Let me see what I can do about it."

Ernie reverses course, walking in his original direction, thereby moving his plane of simultaneity to Andrea Future.

"I fixed it," she informs him. "The ships aren't going to Earth."

"What'd you do?"

"Convinced them to pay a visit to the Cigar Galaxy first. They need sorting out more than you folks in the Milky Way. It will take them about eleven million years to get that one knocked into shape. You're safe for now."

Thanks, Hon. You rock!"

"Any time," says Andrea Future. "No pun intended." And rings off.

Simultaneity of Events vs. Photons

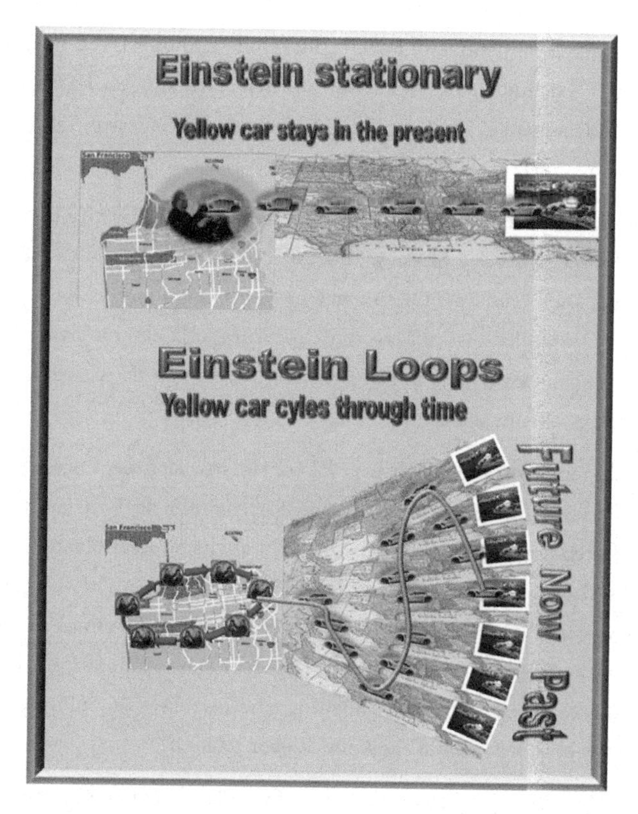

However, you may have noticed some caveats with the time-leveraging story of the Andromeda Paradox. The primary one being that to enable it, Ernie and Andrea would have to communicate instantaneously across millions of light years via their Cosmic Phone, thereby violating all known laws of physics. Even if they could somehow get connected, there are many other motions of the Earth, Sun, and the galaxies through space, causing the planes of simultaneity to ratchet back and forth in a multiplicity of ways.

Even if their Cosmic Phones were somehow programmed to cancel all other motions except theirs, they'd still be bypassing the Relativistic process of forcing the simultaneity plane of an event forward or backward by motion. Ernie would be moving through space relative to Andrea, but the Cosmic Phone communications between them would be moving through some channel outside space that wouldn't be using Relativity of Simultaneity to adjust their times of origin to enforce a constant speed of light. Ernie might be talking to her on his timeline as if she was sitting next to him.

Yet another difficulty is the motion of Ernie walking back and forth on Earth would have to be magnified into large displacements of time on Andromeda when leveraged by 2.5 million lightyears. Real physics does not work this way. It works by leveraging large distances into smaller ones; for example, by using 10 pounds of force to move one end of a lever 10 feet downward, to lift 100 pounds of weight 1 foot higher at the other end. Ernie walking on Earth would be unlikely to create a large time displacement 2.5 million light years away. Physics does not magnify effects as the distance from their source increases.

If these difficulties could somehow be surmounted, and simultaneity planes harnessed to link the past and future, we'd still have to know how to resolve the ubiquitous time travel paradoxes. How could Ernie affect such monumental changes in Andromeda merely by walking around with a phone on Earth? When he walks toward her, Andrea Future tells him an invasion fleet is on the way to Earth. When he walks backward, he informs Andrea Past about it, prompting her to convince the Galactic Council to forego the invasion.

When the decision to invade Earth is revoked, the invasion fleet presumably dematerializes from its Earthbound course and rematerializes on its Andromeda base as if it never left. What happens to all the photons released into space when the invasion fleet blasted off? Do they all jump back into the atoms they were emitted from? What happens to all the atoms in the brains of Andromedins who were on the invasion fleet or saw it take off? Do the atoms in their brains get reset to lose the memory of the fleet blasting off to attack Earth? Does Andrea Future lose her memory of the fleet taking off as soon as she tells Ernie about it, and he relays the information to Andrea Past? If she loses her memory of it, how would she know it ever took off?

Likewise, with the warning of the storm over Indy. The warning from the future is sent when it is known the city is underwater and many have drowned. When the warning is received in the past, people heed it and flee. Do the bodies slain by the future flood rise from their watery graves?

These so-called paradoxes give us some insights into the nature of time. Tens of thousands of books and hundreds of millions of words have been written about it, beginning with ancient philosophers like Plato and St. Augustine and continuing through contemporary ones. My definition is:

"Time is the observed effect of forces moving through space at the speed of light."

Gravity, propagating at the speed of light, tugs the Sun, Moon, and planets along in their orbits, giving us the astronomical measure of time. Electromagnetic forces moving at light speed are transmuted into mechanical forces moving the hands of analog clocks or lighting electronic clocks. Chemical reactions animating physical and biological processes are derived from fundamental electromagnetic and nuclear forces moving at the speed of light.

The speed of light and other forces is the speed of time, and the properties of space define both.

Particles emitted by objects make their effects felt on other objects at the speed of light. When a light is turned on, its photons radiate into the Universe, changing the characteristics of every object they encounter. The event is etched in into the fabric of the Universe. Would it be possible to go back in time and block the turning of the light on, putting those quintillions of photons and force-carrying particles bounding across the Universe back in their holsters?

What, then, is the flaw in these simultaneity plane diagrams ubiquitous in the literature of Relativity? I think it is like looking at a road map and imagining that as soon as you start the car and get on the highway, your time of arrival at your destination is predetermined. Spacetime diagrams assume that when an object starts moving, its simultaneity plane is bent across the entire Universe, as if one tiny motion at the start of a journey can be extrapolated to Eternity.

In the chapter heading diagram, Yellow Car leaves the garage in Washington D.C. and travels toward San Francisco. Green Einstein is stationary waiting for it to arrive. It never leaves his "now" moment. If he is in contact with it by radiophone, he and Yellow Car will be on the same timeline.

Red Einstein is moving in a circle tour of San Francisco while Yellow Car is moving toward him. As Red Einstein circles in the direction toward the Yellow Car, its simultaneity plane ratchets up into the future. As Red Einstein changes direction to move away from Yellow Car, its simultaneity plane falls pastward. Is the event of Yellow Car leaving the garage really ratcheting back and forth through time merely because Einstein is driving in a circle? Or is the time stamp on the Yellow Car and not the event of it leaving the garage? If it is only a stamp on Yellow car, then the time of the event that emitted it does not change.

Once Simultaneity is viewed this way as a granular moment-by-moment process that doesn't happen all at once, many seeming paradoxes go way. The Andromeda Paradox is the first to fall. It assumes Ernie leverages a 6 kph speed across the entire 2.5 million light years to Andromeda into 5 days at the Andromeda end merely by taking the first step. The calculation for bending the time between Earth and Andromeda for a person walking 6 kilometers per hour is:

(Walking speed / speed of light) * distance to Andromeda = (6 kilometers per hour / (300,000 km / second for speed of light) * 3,600 seconds per hour)) * 365 days in a year * 2,500,000 light years to Andromeda = 5.069 days into the future if you're moving toward it, or into the past if moving away.

However, when looked at in the granular way of happening moment by moment, the calculation takes on a different hue. Since 6kph is 180,000,000th the speed of light, Ernie would have to walk 180,000,000 seconds, or 5.71 years to build each second of simultaneity differential, thereby having to walk for the whole 2.5 million years to get the 5.069 days of simultaneity difference.

This fallacy of leveraging Relativity of Simultaneity across distance is reminiscent of the fallacy whereby you supposedly send a signal faster than light by traversing a laser pointer at a distant object, for example, across the image of the moon in the sky. When the arc of photons arrives at the moon, they cover a line across the moon's surface instantaneously, sweeping from side to side at multiples of the speed of light. The fallacy, of course, is that a light beam isn't all one thing. It's a stream of photons, each photon pursuing an independent path across space. The width of the photon arc is tiny when emitted from the laser pointer on Earth, the arc increasing with distance as the photons disperse to hit the moon in different places. No information is transferred from one side of the moon to the other instantaneously; all parts of the arc receive the same information at the same time, the way any other signal radiates from a central emitter.

I see simultaneity planes as the same phenomenon --- an agglomeration of individual photons moving independently away from an event occurring at one time and place, not as monolithic planes extending across the Universe. I believe each photon is time-stamped when it is received, according to the relative motion of the sender and receiver during its time in flight, while the time of origin of the event emitting it remains constant. In this example, the time stamp on the hotons changes with the circling motion of Einstein's car, but the time of Yellow car leaving the garage doesn't.

When Yellow Car leaves the garage, it interacts with the broader Universe, spreading hanges to anything it encounters. Once it leaves the garage there is no going back in time to inhibit t from leaving. Likewise, when an event happens in space, it releases photons and force-carrying articles into the Universe, changing the energy and mass states of everything they interact with. Ve call these changes "time" because that's a handy word to describe the cumulative changes to bjects in the Universe as they are impacted by photons and forces. These changes are irreversible, o no going back in time to undo them.

If this idea is correct, going back in time by changing your simultaneity plane would be like hotoshopping a picture of yourself when you were a small child. You can Photoshop the picture to hange the photons you see reflected from the picture now, to make your image more appealing --- erhaps by drawing your face more handsome and your legs longer, and your body thinner. These hanges to a photograph taken decades earlier won't change your real-life appearance now. Likewise, hanging the Relativity of Simultaneity of photons won't enable you to reach out across time and hange or inhibit the event that emitted them.

Time Dilation

Whereas Relativity of Simultaneity is uncertain, time dilation is a demonstrably rea phenomenon with a straightforward explanation. To understand it, let's recognize that light cann be compartmentalized within a contained space the way a propagation phenomenon like sound i Sound is compartmentalized within a car because it travels through air inside the car, moving wit the car. From inside the car, it looks like photons are contained the same way. Photons from a ligh source inside the car will be absorbed by objects inside the car and re-radiated as heat or will b reflected as colors, depending on their wavelengths. Photons will not leave the moving car if it i sealed up tight to have no windows they can escape through. However, light is not reall containerized this way, because it moves through space independently of the car, at a constant spee of c, whether the car is there or not:

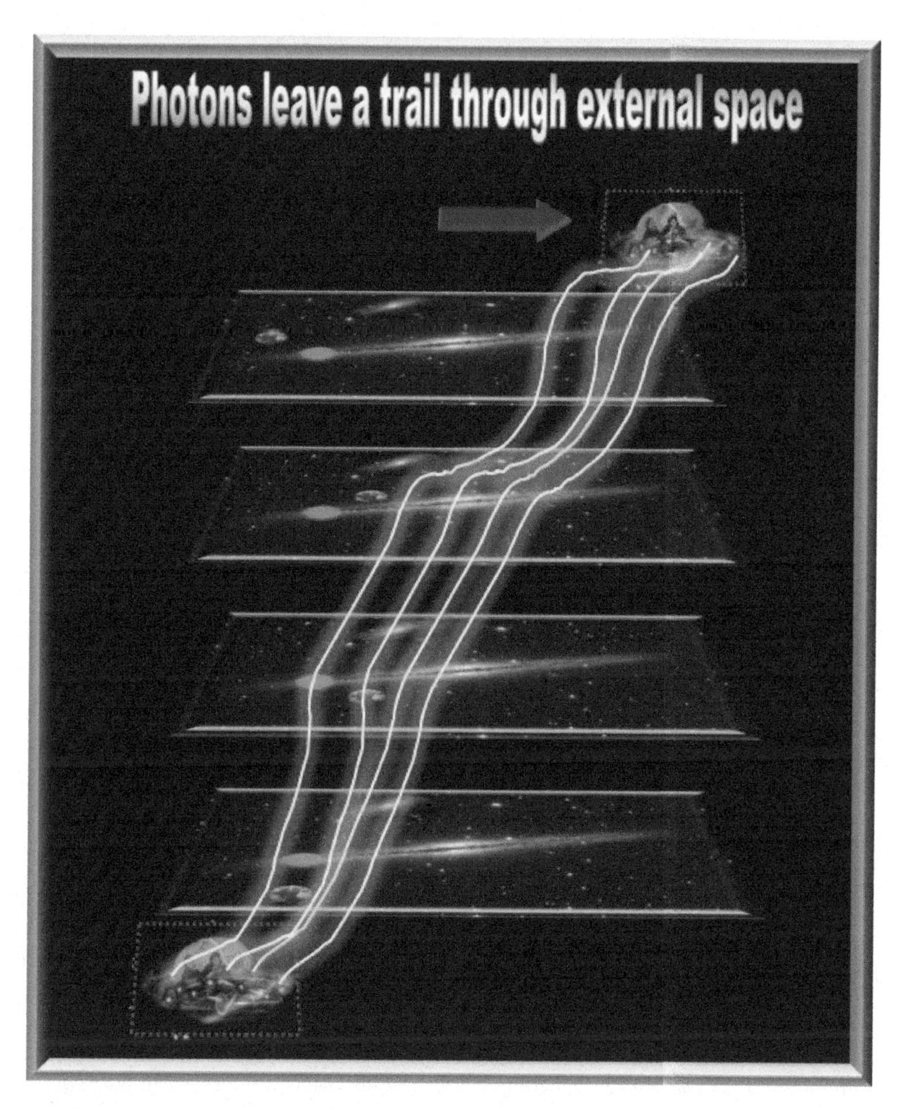

The faster the car travels, the longer its trail through external space, and the slower the photons travel internally inside the car. If the car were to reach 100% light speed, the photons inside would be frozen motionless, their external trails through space using up all their speed, leaving nothing left to move them inside the car. Because photons are the agents of time, time would freeze inside the car. This is what time dilation is, the slowing of time inside a moving object due to its photons and other force-carrying particles being spread across external space by the motion of the object.

Because time is mathematically a 4th dimension perpendicular to all three spatial dimensions, time dilation is a trigonometric function whose effect is leveraged at speeds close to light:

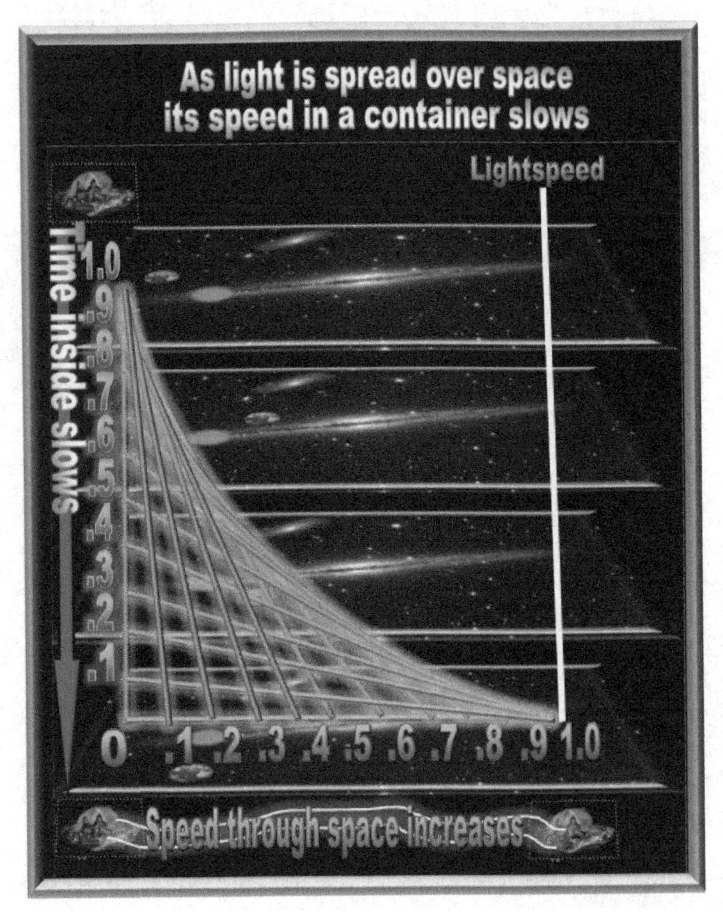

The numbers at the bottom show how close to lightspeed Einstein's car is travelling. As it gets closer to 1.0, it pulls the time duration arrow down faster and faster toward zero elapsed time. Here it is, more clearly:

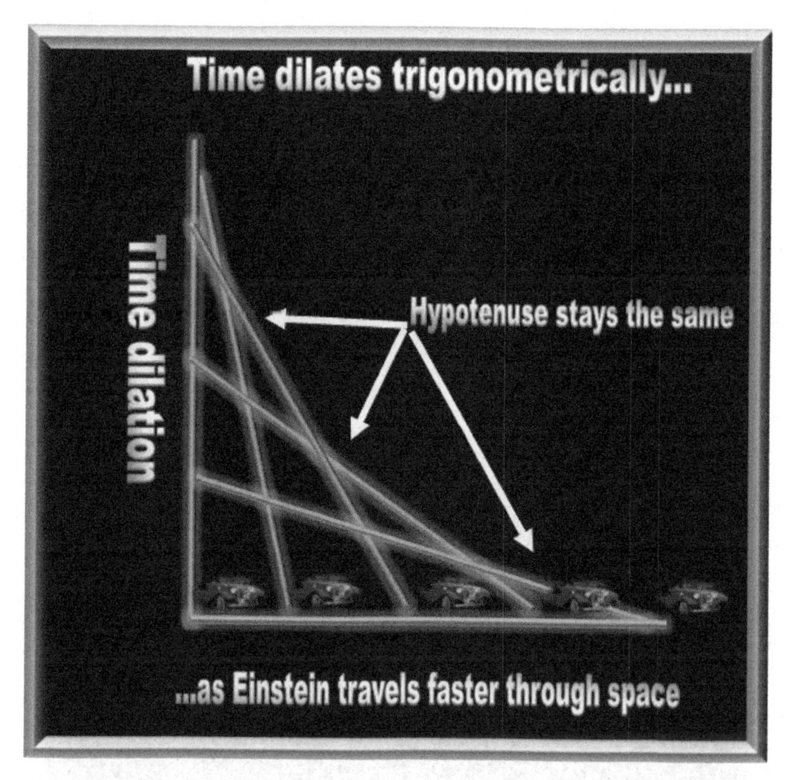

The mathematical construct is a hyperbola whose hypotenuse remains of fixed length as it slides up and down the vertical and horizontal axes, the horizonal axis representing all three dimensions of space and the vertical axis the dimension of time. The drawing by Hermann Minkowski, another of Einstein's professors, shows its mathematical elegance.

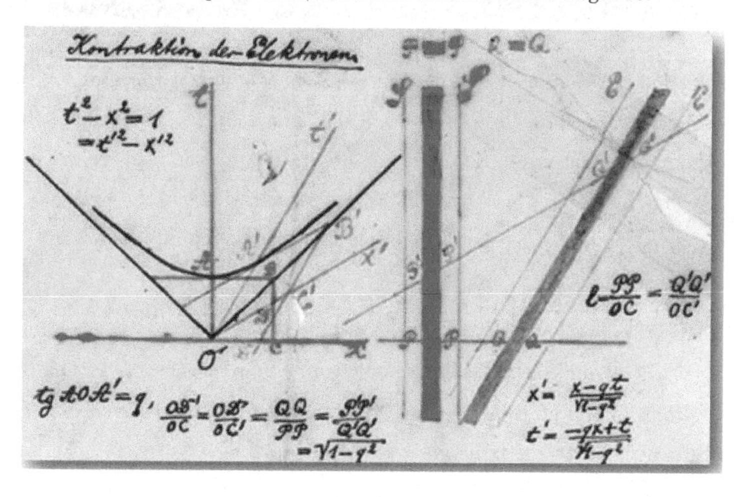

Macrocosm #3 Andromeda Expressway!

As a trigonometric function, time dilation requires prodigious energies to accelerate objects near the speed of light. If we could somehow obtain that much energy, we could go at least as far as the Andromeda Galaxy and return to Earth in one lifetime. The energy input would require cosmic engineering --- perhaps building a laser lens as big as the sun to focus the bulk of its energy into a beam to transmit power to a trans-galactic spaceship, while having another star put in harness at the destination to decelerate it to a stop.

If we had this cosmic source power, we could accelerate a ship comfortably at 1g (Earth gravity) for 14.3 years of dilated on-board travel time, then decelerate it for 14.3 years as it approaches its destination and be in Andromeda. We might spend 20 years checking out a planet or two there, then return to Earth. We'd be gone 28.6 + 20 + 28.6 = 77.2 years of dilated and normal ime. Time elapsed on Earth would be 2.5 million years (approximately) + 20 years + another 2.5 nillion years. Here are couple programs that calculate how long it would take you to go to Andromeda, or anywhere else in the Universe, at 1g acceleration followed by 1g deceleration:

https://www.omnicalculator.com/physics/space-travel

https://spacetravel.simhub.online/

Our top speed at the midpoint when acceleration is at its maximum would be 9999999999996994% the speed of light. The average time dilation on the ship would make you think you were covering 2,500,000 light years / 28.6 years dilated time = 87,412 times the speed of light. Each second you experience on the spaceship would be more than a day of elapsed time on Earth. Each day would equal 239 years of Earth time. If you looked out a porthole, you'd see light passing at he same speed as it did on Earth, because light travels the same speed for dilated time as for at-rest ime. Because you cover much more distance per second of dilated time, you'd think the distance between Earth and Andromeda had shrunk from 2,500,000 years to 28.6 light years. The apparent and real effects of what you'd see are shown in this Youtube video:

https://www.youtube.com/watch?v=vFNgd3pitAI

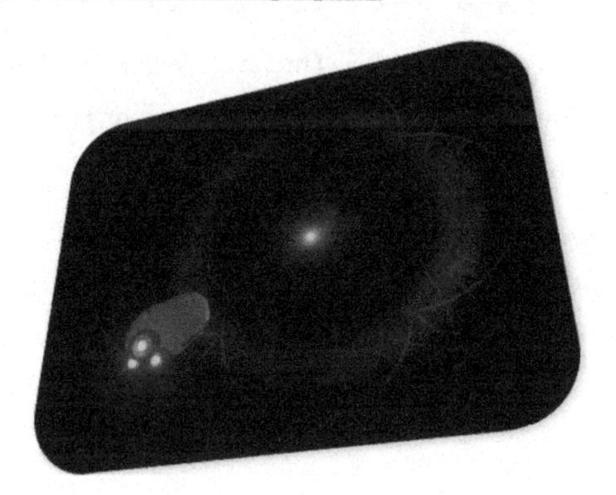

The Duality of Spacetime

In Relativity, objects composing the Universe swim in the sea of "spacetime," the mathematical union of space and time, invented by Henri Poincaré and Hermann Minkowski to describe the effects of motion through space:

The views of space and time which I wish to lay before you have sprung from the soil of experimental physics, and therein lies their strength. They are radical. Henceforth space by itself, and time by itself, are doomed to fade away into mere shadows, and only a kind of union of the two will preserve an independent reality.

- Hermann Minkowski

Time becomes a mathematical 4th dimension when calculating time dilation as a trigonometric function of the three dimensions of space moving perpendicular through time. Time is also carried through the three dimensions of space by photons and force-carrying particles imparting a linear mathematical function to the Relativity of Simultaneity. This trigonometric and linear character reflects the dual nature of space and time both as a three- dimensional and four dimensional entity. The four-dimensional nature of spacetime has space encapsulated in its dimensions, linked by the perpendicular dimension of time:

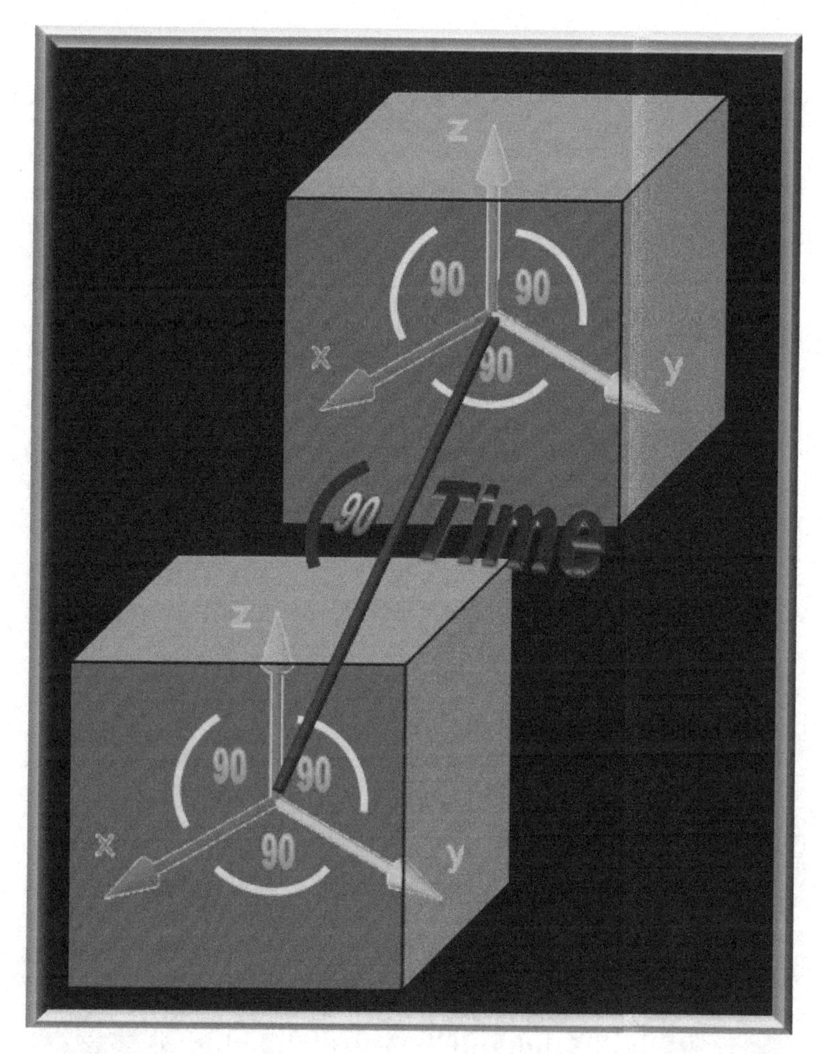

Time is not a true dimension one can travel through at will, like the three physical dimensions. It is a mathematical dimension traversed according to each object's movement through space, such that Relativity of Simultaneity hypothetically moves the time of origin of an event backward and forward through time, as observed by other moving objects, while time dilation is a proven effect of stretching a light trail across space.

Spacetime is usually described as the 4-D incarnation of space evolving through time. But time is also an integral component of space, being transmitted through space by photons carrying electromagnetic energy and other force-carrying particles, including bosons and gluons, carrying the nuclear forces. It is just as true to say that time lives within space as it is to say that space lives within time:

I view this question about whether time dominates space or space dominates time as similar to the question of whether water surrounds land or land surrounds water. If you're looking at a globe, you see North America as an island continent surrounded by oceans and a tiny strip of water in the Panama Canal. If you're looking at Lake Michigan inside the North American landmass, you see a body of water sounded by land. If you look at Beaver Island in Lake Michigan, you see land surrounded by water. If you look at Font Lake in Beaver Island, it's water surrounded by land. And if you look at the islands in Font Lake, it's land surrounded by water. It all depends on what you're looking at. If you wanted a generic term for both, you might call it "waterland" or "landwater."

The Lorentz Transformation

Hendrik Lorentz, winner of the Nobel Prize in Physics in 1902, was Albert Einstein's professor. He deduced the constant speed of light and devised his namesake Lorentz Transformation as the mathematical formula describing light's constant speed of c in all circumstances of motion between objects emitting photons and force-carrying particles, and objects receiving them, regardless of motions of either or both while the photons and force-carrying photons are in flight. The Lorentz Transformation consists of the two time-bending functions we have studied:

1) Relativity of Simultaneity, a linear function, hypothetically advancing or regressing the time of an origin of an event according to how much movement there is between the object transmitting photons and particles of force and the objects receiving them while the photons and particles are in flight. Moving away from an object supposedly causes the time of origin to happen later, while moving toward an object causes its time or origin to happen earlier.

2) Time dilation is the trigonometric Gamma Function representing time as a 4th dimension perpendicular to the three dimensions of space. Travelling at 10% lightspeed makes time pass about .5% (one half of one per cent more slowly) relative to objects at rest; travelling at 50% lightspeed makes time pass about 15.5% more slowly; travelling at 86.6% lightspeed makes it pass about 50% more slowly; and travelling at 99% of lightspeed makes time pass about seven times more slowly.

Let's combine Relativity of Simultaneity with Gamma to create the Lorentz Transformations or the car-centric examples I've used for stationary and moving Einsteins. The top of the first picture below is Relativity of Simultaneity only for Einstein moving at 30mph (half our car-centric speed of light), while the lower half combines the Gamma of time dilation into the complete Lorentz Transformation:

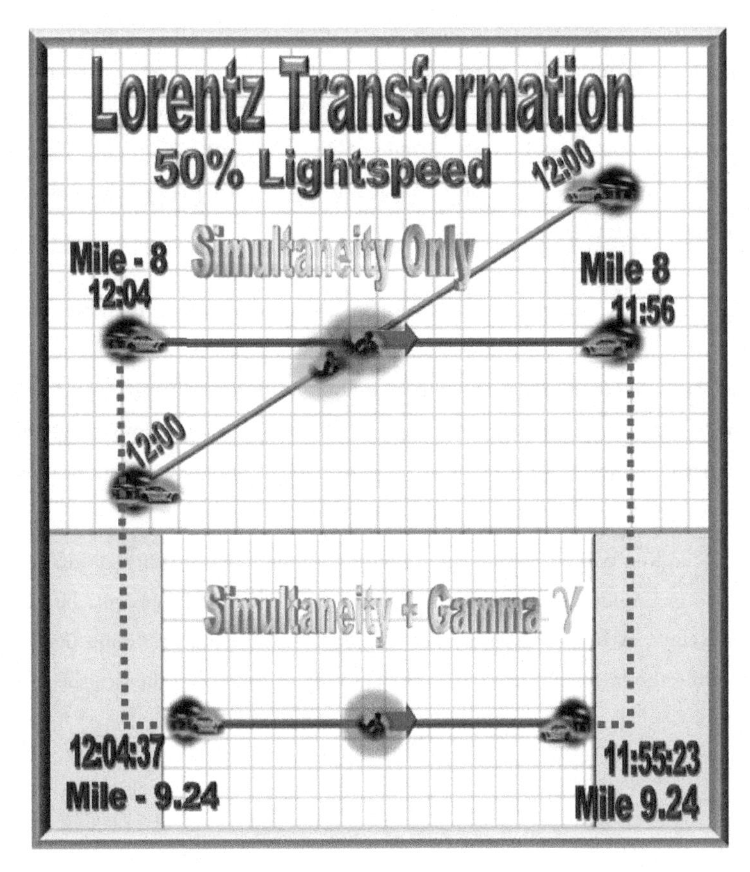

The Gamma Function accounts for the time dilation of Red Einstein. At half the speed c light, the trigonometric function of Gamma extends each of Red Einstein's seconds to approximatel 1.155 of stationary Green Einstein's. Light, having a constant speed, must show the same pat through space for Green and Red Einsteins. They will both see light covering 300,000 km/sec in thei own frames. However, because Red Einstein's dilated seconds are longer than Green Einstein' seconds, the distance light covers in his seconds must also be longer --- in the same way extrem time dilation covers 2,500,000 light years of distance in 28.6 years of dilated time in Andromed Expressway!

Red Einstein will see the gamma multiplier added to his Relativity of Simultaneity of tim and distance, such that the 8 miles are multiplied by 1.155 to become 9.24 miles, while the 4 minute of simultaneity are multiplied by 1.155 to receive another 37 seconds before and after.

At 95% the speed of light, as shown next, the Simultaneity has increased from 8 minutes 50% the speed of light = 4 minutes to 8 minutes * 95% the speed of light = 7.60 minutes. The Gamm multiplier expands to 3.20, so the time advances and regressions for Simultaneity * Gamma are now 7 minutes and 36 seconds * 3.20 Gamma = 24 minutes and 20 seconds:

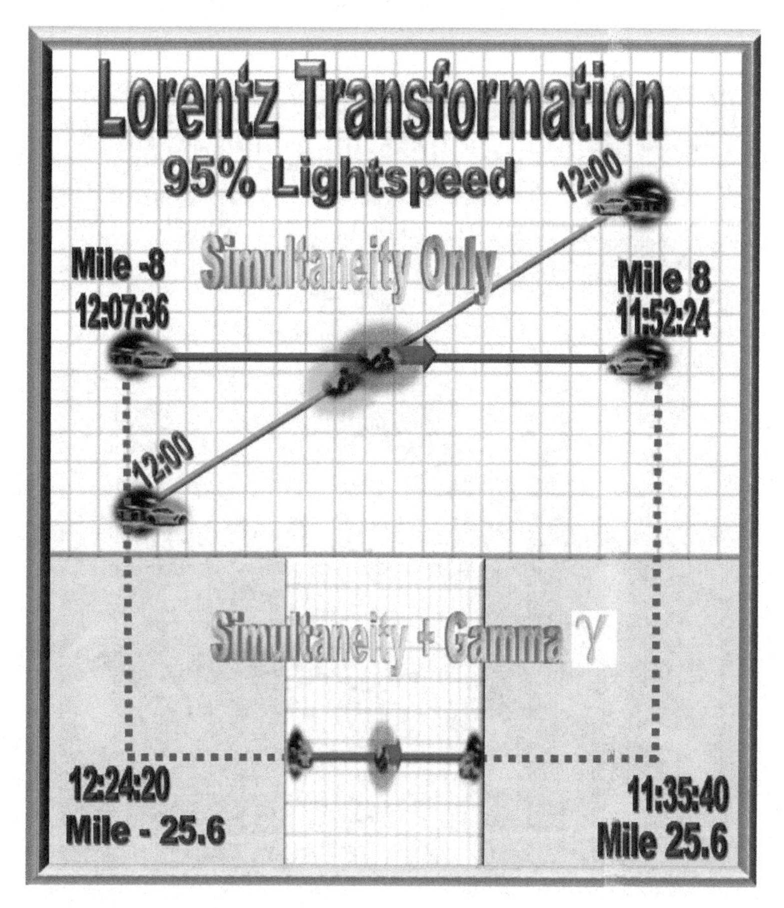

The more you approach the speed of light, the thinner the Universe appears, but the more fantastic are the times and distances that light has to cover at its constant speed of *c* through a shrinking Universe, according to your time-dilated perspective. As with simultaneity planes, I am not convinced that Gamma shapeshifts the Universe every time you change your velocity. Perhaps the constant speed of light, acting through photons and force-carrying particles, gives the appearance that it does by acting on the particles, not the event that spawned them.

Macrocosm #4 The Twins Paradox

The most ubiquitous discussion in Relativity revolves around the so-called "Twins Paradox" of determining why an Earthbound twin ages faster than a twin who travels at near-light speed to a distant place and returns to Earth. If all motion is relative, and there is no privileged frame of motion, why can't we say it's the Earthbound twin who moves through space and stays younger while the travelling twin remains stationary and ages normally?

The "pat" answer is that the travelling twin experiences acceleration, deceleration, and a change of direction that sets him/her apart from the idling Earthbound twin when they meet up again. When this explanation is challenged, the metaphorical twins may proliferate, becoming attached to Earth and various incarnations of spaceships plying between Earth and distant worlds, each twin trying to nail down the precise moment when they, and not their Earthbound twin, are assigned the time dilation that keeps them young. The explanations of why the travelling twins receive the time dilation turn out not to be as simple as they are billed.

The standard explanation comes in two variations. The most popular one shows the ship travelling far from Earth at high speed, then reversing course and heading back to Earth, supposedly causing the ship and its occupants to experience a "simultaneity gap" in evading some of the time the Earthbound twin experiences:

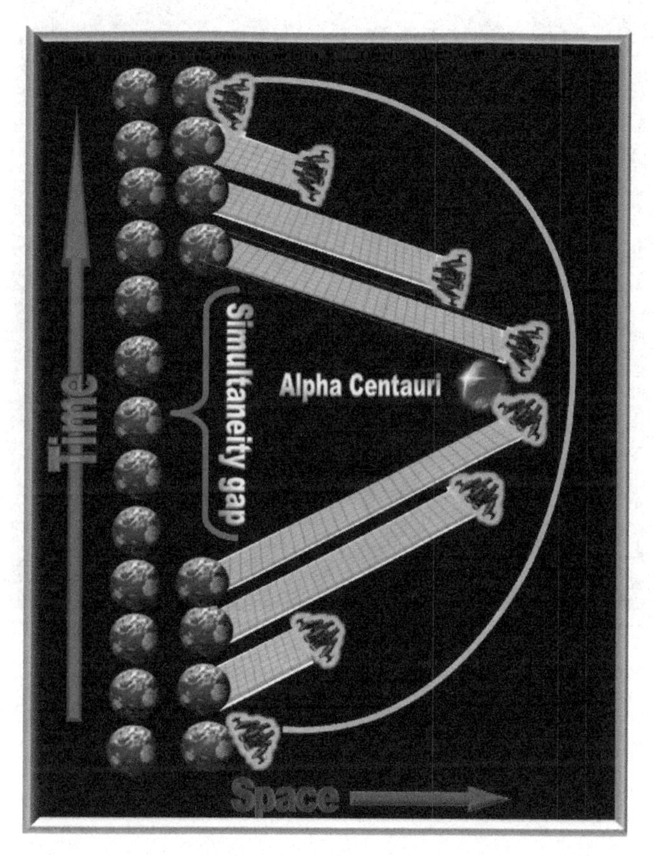

The second variation on this idea is the light-burst or videogram theory, whereby periodic light bursts --- or better yet, real-time webcam videos --- are transmitted between the travelling twin and the Earthbound twin allowing each to see how much the other ages moment by moment, after accounting for the propagation delay it takes light to get from Earth to the ship, and vice versa. It is alleged that the light bursts or videograms appear symmetric to the Earthbound twin, but are asymmetric for the travelling twin, due to his/her reversing direction, and this supposedly accounts for the slower aging:

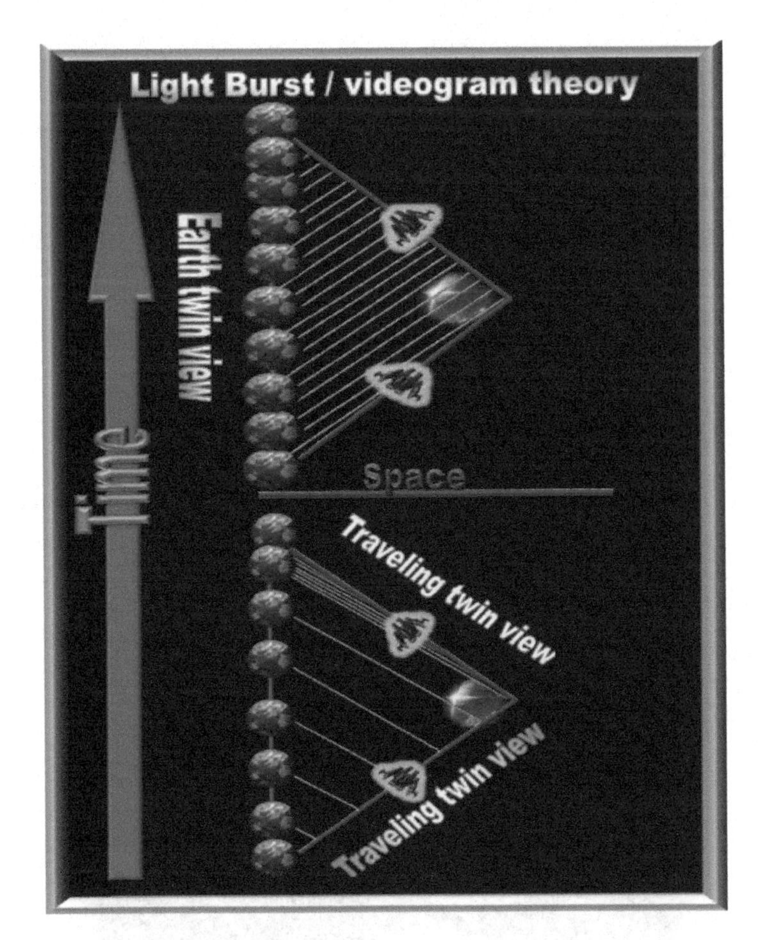

These theories are sometimes challenged. For example, if it is a change in direction after a long period of high speed motion that sets the time dilation for the travelling twin, then why couldn't we consider the ship to be rotating around a stationary point in space, while the Earth twin travels a long arc of changing direction around it, thereby being assigned the time dilation:

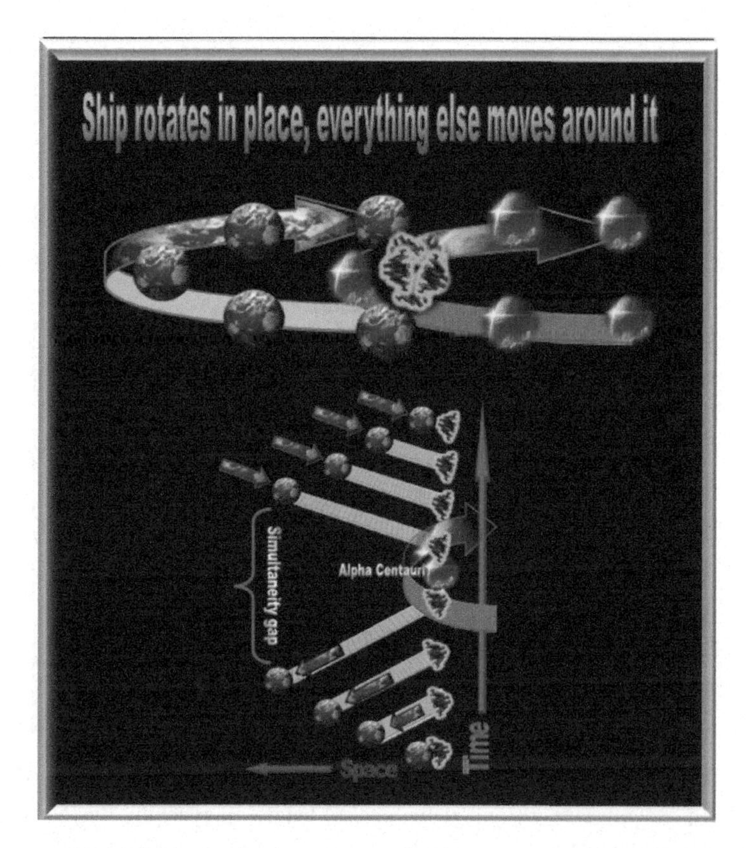

"Because only the ship experiences a change in direction caused by the application of a force om the rocket engines," comes the reply. Ok, so now let's take the change in direction out of the icture by letting the ship land on a planet and never returning to Earth. What if they exchange ideograms as soon as the ship lands? Who will get the time dilation then? You'd think it would be ie travelling twin because they covered a larger distance. We could also argue that the symmetry as broken by the acceleration and deceleration of the travelling twin, even if there's no longer a iange in direction:

Now suppose we remove acceleration, as well as change of direction, from the question b presuming the ship was already moving at high speed before approaching Earth. Maybe it was a alien ship accelerated millennia ago and now passing through our solar system as a way station extract energy from our Sun before moving on to another galaxy. Science Fiction legend Arthur C Clarke wrote a fascinating story of this scenario in his book ***Rendezvous with Rama***. If the shi passes Earth and Alpha Centauri in a straight line at high speed, with no change in speed c direction, would the creatures aboard not experience time dilation relative to Earth during th interval between those points?

Relativity theorists are prone to saying, "No, they would not, because there is no privileged frame of reference for objects in inertial motion. If the ship experiences no acceleration or change of direction between Earth and Alpha Centauri, it cannot experience time dilation relative to anybody on Earth." However, there are straight-line phenomena known to experience time dilation:

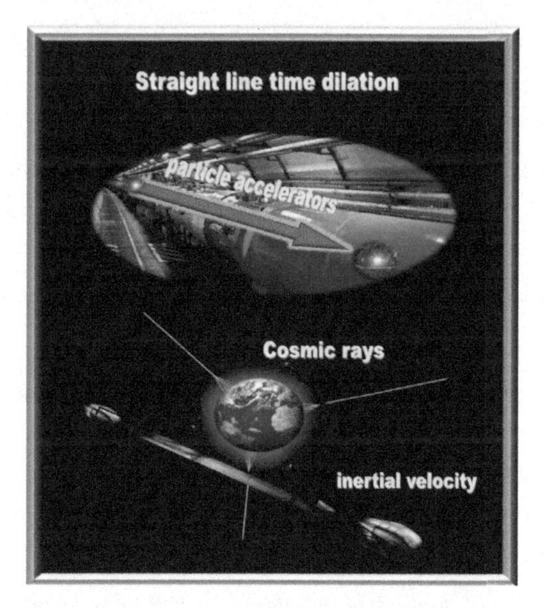

Particle accelerators not only accelerate existing particles, but also create new particles from energies released by colliding particles. The newly created particles, born with high speed, have time dilated extended lifetimes relative to slower particles. A similar natural phenomenon occurs in cosmic rays, which are fast-moving protons and neutrons ejected from stars. When the particles enter our atmosphere, they collide with atoms and produce short-lived particles called muons that would normally decay before they reach the surface. But the fast-moving muons are time-dilated to about 24x their normal lifetimes, thus living long enough to be detected at the surface. Granted, these are tiny particles created spontaneously in high-energy collisions, but they do unequivocally receive time dilation moving in a straight line without acceleration. I have never received a satisfactory answer of why this happens from the "acceleration and change of direction" mavens.

Nevertheless, this idea that an object in constant straight-line motion might be differentiated from any other object in constant straight line motion, refutes the key dogma of Relativity that "there is no privileged inertial frame of reference," meaning that if an object is not changing speed or direction, it may be considered at rest relative to the rest of the Universe, and the laws of physics apply consistently to it, such that there is no way to determine whether it is intrinsically at rest or in motion, or experiencing time dilation:

It feels like we are standing still, even though we know we are moving in orbit around the sun, because there is no way, not even in principle, of deciding what is standing still and what is moving.....And if all you can do is speak of how the book moves relative to you as you sit in your aircraft seat, or relative to the ground, or relative to the sun, or relative to the Milky Way, but always relative to something, then absolute motion is a redundant concept.

...The key point that underpins this entire book and forms one of the very cornerstones of modern physics is that, provided the aircraft is not accelerating or decelerating, none of these experiments will reveal that we are in motion. Even looking out the window doesn't tell us this, because it is equally correct to say that the ground is flying past beneath us at six hundred miles per hour and that we are standing still.

Cox, Brian; Forshaw, Jeff. Why Does E=mc2? (And Why Should We Care?). Da Capo Press. Kindle Edition.

Indeed, I have used this principle to explain the Relativity of Simultaneity by assuming that whenever Einstein is moving at constant speed, he is at rest and the Universe is moving around him. It is a sound principle drilled into the bedrock of Relativity theory. But is it the final word? Einstein's revered contemporary Ernst Mach, renowned in many fields of physics, from his eponymous sound barrier scale to cosmology, thought it might not be.

'The whole direction of thinking of this theory [of Relativity] is in concordance with that of Mach, so that it is justified to consider Mach as the precursor of the general theory of relativity,' [Einstein] wrote in 1930. In the last interview given by Einstein, two weeks before his death, he reminisced with evident pleasure about the one visit he had paid to Mach and he spoke of four people he admired: Newton, Lorentz, Planck, and Mach. They, and Maxwell, and no others, are the only ones Einstein ever accepted as his true precursors.

Pais, Abraham. Subtle is the Lord (p. 283). OUP Oxford. Kindle Edition.

Mach was both a progenitor and a skeptic of Relativity, so perhaps his ideas have merit in fleshing out some of the unsettled controversies of Relativity, like the Twins Paradox.

Relativity vs. the Machian Universe

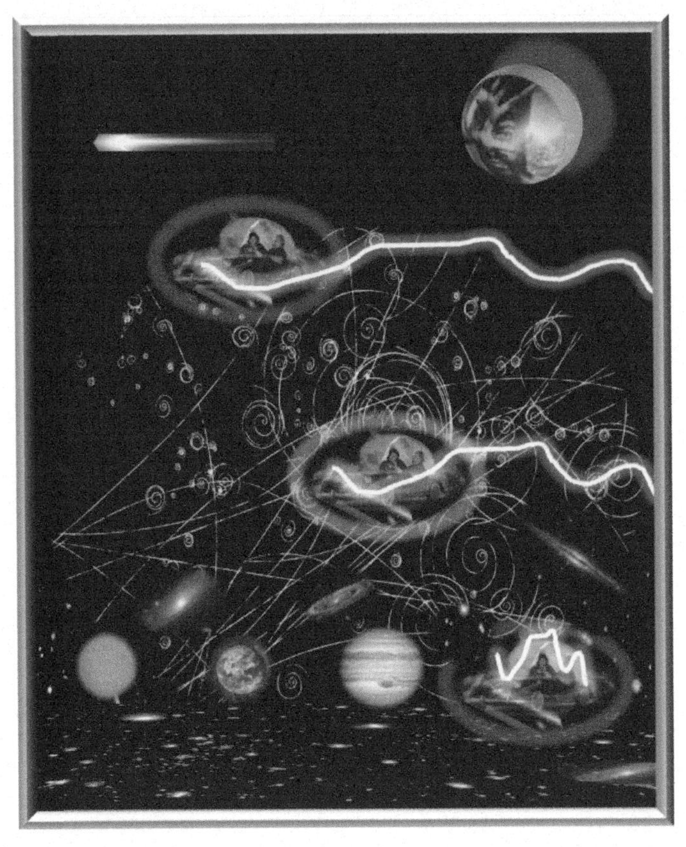

The concept of a baseline velocity in the Universe --- a privileged frame of reference by whic[h] all others are measured --- goes so much against the grain of Relativity that to suggest there migh[t] be one has been called "a crime against science." Each object considers its own light path to b[e] internalized as a stationary point in space, and every other object's light path to be longer an[d] therefore time dilated. The Einsteins and The Observer both consider themselves stationary whil[e] the other moves:

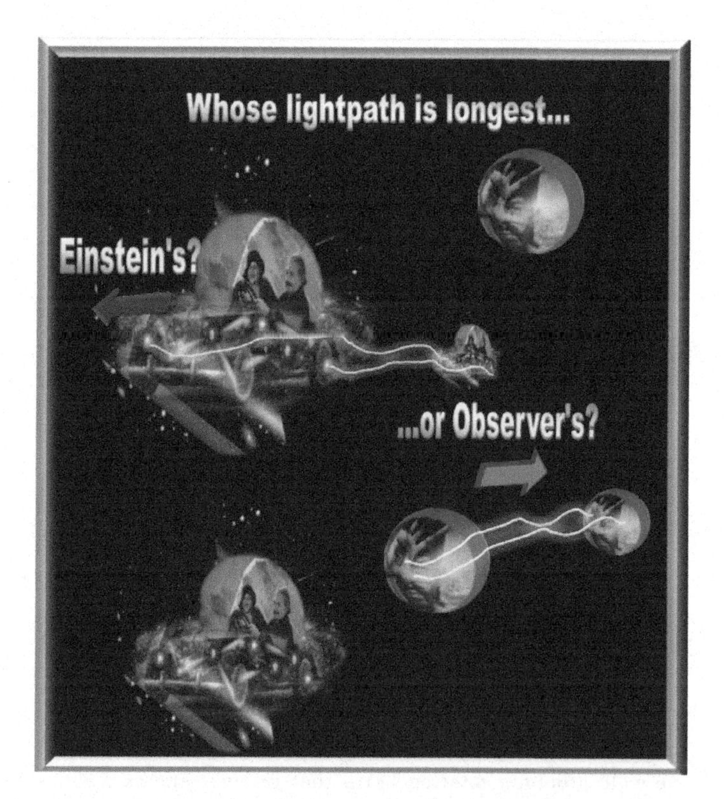

And who's to say that either the Einsteins or the Observer has the definitive view of their light trail? We could layer on another observer who sees both differently than they view themselves:

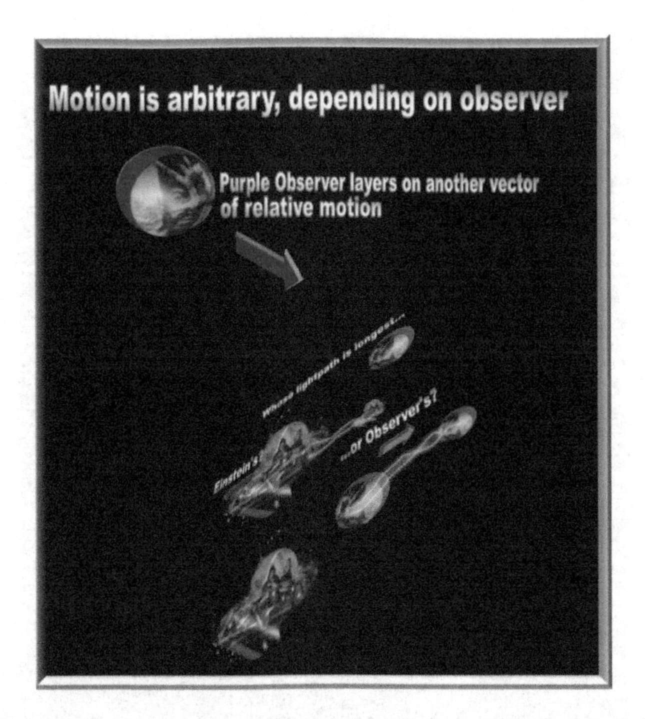

We could layer other views on top of that, making it impossible to say who's moving and who isn't. In real-life, we walk around a rotating Earth that revolves around a Sun orbiting the Milky Way Galaxy that moves through space with a Local Group of Galaxies, moving with larger galactic structures that move with imparted motion from the expanding Universe. Who knows what other hidden motions are behind all that? It only makes sense to say that every non-accelerating object is stationary to the rest of the Universe. Indeed, this is the view I have taken in explaining the Relativity of Simultaneity.

And yet, the complexity of Relativistic lore devised to explain the Twins Paradox leads me (and some others, especially Mach himself) to believe there is something intrinsic about high velocity that causes time dilation even when an object moves at constant speed in a straight line. If a spaceship accelerates from Earth at near-light speed and travels to Andromeda, I believe time dilation is experienced while the ship is travelling at constant speed, not when it accelerates, decelerates, or turns around --- the same straight-line phenomenon as happens when cosmic rays impact our atmosphere.

Our spaceship to Andromeda might have accelerated prior to passing Earth in the ***Rendezvous with Rama*** scenario. Whether it accelerated before reaching Earth or not, I believe the occupants will experience time dilation the whole way from Earth to Andromeda, aging only 28.6 years in the Andromeda Expressway! scenario, while Earthlings age 2.5 million. If the ship was already travelling near light speed when it passed Earth, it would have had to accelerate at some prior time that could have been the recent past, or as far back as The Big Bang 14 billion years ago. Perhaps velocity relative to the point where the Big Bang originated might be the at-rest baseline velocity standard for all objects in the Universe, because that's where the Universe began, at a single point with zero velocity relative to anything else. If that be so, then any object's motion, and therefore its time dilation, could be calibrated according to its velocity relative to the point of origin of The Big Bang.

This idea of a privileged frame of reference based on an at rest center of the Universe was invented by Ernst Mach and studied by Einstein as he formulated Theory of Relativity. Einstein's biographer Abraham Pais writes:

Enter Mach. In this note Einstein declared, 'This [conclusion by Mach] lends plausibility to the conjecture that the total inertia of a mass point is an effect due to the presence of all other masses, due to a sort of interaction with the latter.... This is just the point of view asserted by Mach in his penetrating investigations on this subject.'

From that time on, similar references to Mach are recurrent. In the Einstein–Grossmann paper we read of 'Mach's bold idea that inertia originates in the interaction of [a given] mass point with all other [masses]."

In June 1913, Einstein wrote to Mach about the induction effect as well as about the bending of light, adding that, if these effects were found, it would be 'a brilliant confirmation of your ingenious investigations on the foundations of mechanics' [E38]. In his Vienna lecture given in the fall of 1913, Einstein referred again to Mach's view of inertia and named it 'the hypothesis of the relativity of inertia' [E39]. He mentioned neither this hypothesis nor the problem of inertia in any of his subsequent articles until February 1917, when he submitted a paper [E40] which once again marks the beginning of a new chapter in physics: general relativistic cosmology.

Pais, Abraham. Subtle is the Lord (p. 285). OUP Oxford. Kindle Edition.

Today, we can see the fallout from the Big Bang as the Cosmic Microwave Background (CMB) radiation. We have identified its point in space of its origin, known as the "dipole." Any object's velocity relative to that point might be considered its intrinsic velocity. The Sun's intrinsic velocity relative to the CMB dipole is said to be 369 kilometers per second, or about .123% the 300,000 km/sec speed of light. The compounded +/- velocity of all known stars and galaxies out the range of Andromeda 2.5 million light years away relative to our sun is less than 1,000 km/sec relative to the CMB (and also our solar system), telling us that everything near us in space is creeping along slowly on the Universe's speedometer. The Machian view is that anything travelling at near-light speed in relation to the CMB dipole (or our solar system) is experiencing significant time dilation, whether it accelerates, changes direction, or continues in a straight line. If that's so, then the Twins Paradox resolves without all the complexity Relativity theorists layer on, because the twin with the longest light trail relative to the CMB dipole is the one who gets the time dilation, regardless of whether they stop, turn around, or just keep going in a straight line.

Another quibble with Relativity dogma is:

...The key point that underpins this entire book and forms one of the very cornerstones of modern physics is that, provided the aircraft is not accelerating or decelerating, none of these experiments will reveal that we are in motion.

But there are cues as to who is moving. For example, objects moving at high speeds relative to other objects have more mass, resulting from a higher kinetic energy content. When a small object accelerates relative to the rest of the Universe, the Universe does not gain mass, only the object that is accelerated. The object that begins an "inertial frame" of constant velocity after acceleration has gained mass relative to the rest of the Universe, but the rest of the Universe has not gained mass relative to it.

Also, the speed of light is constant, but its frequency downshifts to red and loses energy as you move away from its source, while intensifying toward the blue end of the spectrum while you're moving toward it, because Relativity of Simultaneity impels a Doppler shift to light's wavelength frequencies. If you are moving near lightspeed, the light approaching you from the front will upshift to gamma rays and kill you if you are unshielded, while the photons coming from behind you are downshifted to red. Only the photons from objects moving at the same speed along with you don't show a red or blue shift. If you can see outside your spaceship, the red or blue shift from the ambient light of the Universe gives visual cues as to your speed and direction relative to everything else:

There is also a series of terrestrial thought experiments conceived by Einstein, known as the "train and embankment," regarded as the bedrock of Relativistic phenomenon. I will customize them a bit, in keeping this exposition consistent with its car-centric nature.

Let's say Albert and Elsa are driving long distance. Night settles in, and Albert is napping while Elsa takes the wheel. Albert is jolted awake by simultaneous lighting flashes hitting eight miles behind him and eight miles ahead. The flashes don't last long enough for Einstein to tell if the car is moving compared to the landscape.

Of course, he could just ask Elsa whether she is sleeping or driving, but being a Relativity physicist, that would spoil his fun. He wants to determine his motion relative to the lightning strikes by examining the characteristics of the light flashes. The fact that they arrived simultaneously doesn't tell him anything about his motion, because he knows light is a Simultaneity phenomenon, whose time of origin varies according to the receiving object's motion as well as its own. He recognizes that deducing whether he's at rest or in motion relative to their impact points will be a tough call, since the bolts could have arrived simultaneously if he were stationary, or might have arrived non-simultaneously for stationary observers, while adjusted to simultaneity if Elsa was driving when they hit:

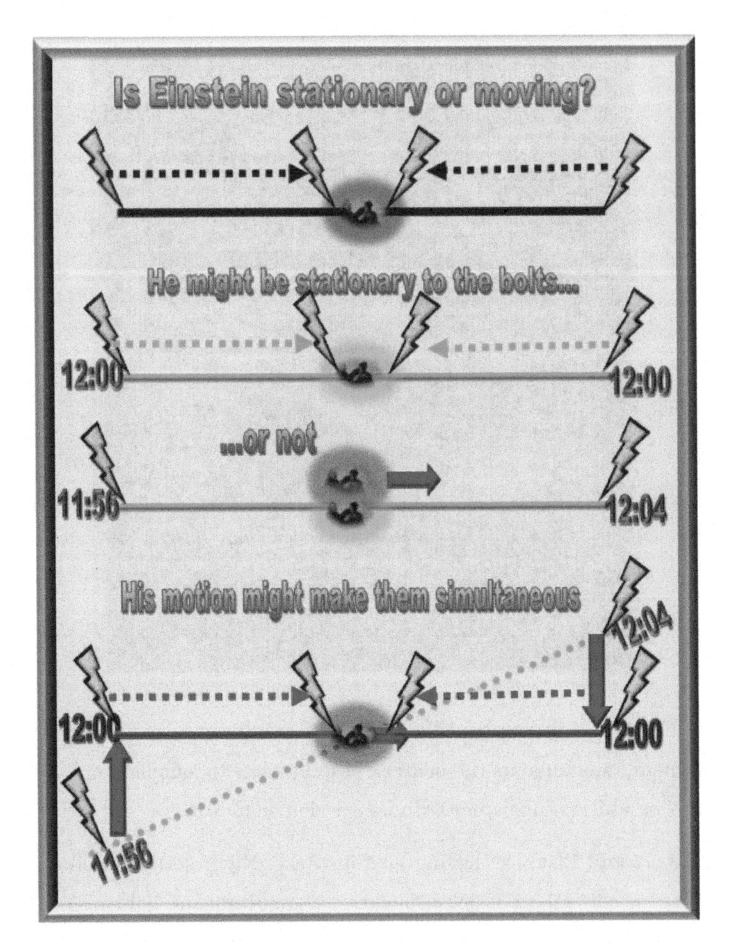

However, Einstein knows that if the bolts are simultaneous because their times of origin have been adjusted by Einstein's motion, the one from behind will be red-shifted, while the one in front will be blue-shifted. If the bolts are simultaneous without being red-shifted or blue-shifted, then Elsa has been napping and the car is stationary:

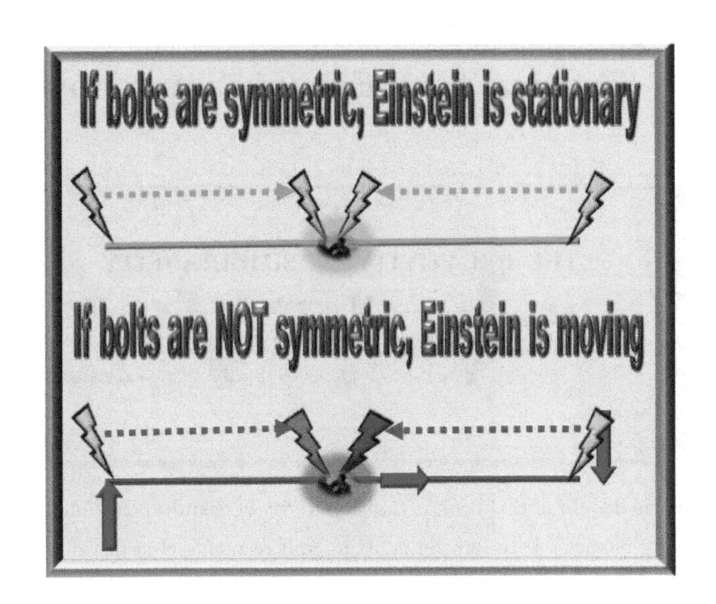

Thus, Relativity may not be as impartial to different states of inertial motion as it is made ut to be. As Mach told us, we cannot really isolate the motion of any object from the rest of the Jniverse, and we cannot deny the effects of the red and blue shifts of light in identifying the velocity nd direction we are moving. If you're the one moving fastest relative to the CMB dipole, you' maybe he one who gets the time dilation relative to someone who's going slower. If that theory has any nerit, then it validates the Machian view of how time dilation is assigned.

I don't want to overdo the Machian Universe by presenting it as an alternative explanation o Relativity. However, some controversies that acquire great complexity in Relativity, such as the Twins Paradox, resolve if we accept the idea that motion is intrinsic as well as relative. Just as the Andromeda Paradox was resolved by understanding that Relativity of Simultaneity is a granular rocess created incrementally by motion, instead of a monolithic simultaneity plane extending across he Universe the instant you begin to move. On the other hand, most Relativity physicists are damant that there is no intrinsic standard for calibrating absolute motion, so it's best to hedge your ets on that one, whatever your view of it. When pressed about Relativistic vs. Machian motion, the nore open-minded physicists are prone to saying, "I think so, hypothetically," rather than, "I know...."

The view from Einstein's Auto Train

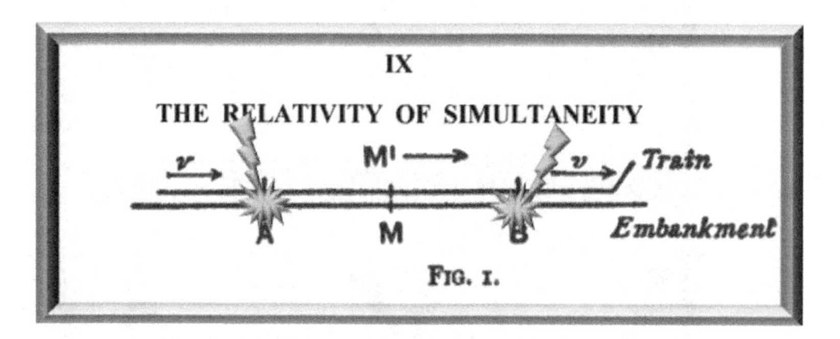

FIG. 1.

The unique insight of this book is that Relativity of Simultaneity follows the flight of photons and force-carrying particles between the emitting and receiving objects, adjusting the time of origin of the emitting event according to the each receiving object's view of the relative motion occurring between emission and reception. This insight debunks the misapprehension that simultaneity planes are monolithic, time-bending structures that materialize across the Universe the instant you begin to move. It resolves the so-called Andromeda Paradox of imagining that small motions on Earth are amplified across the Universe to bend the timelines of distant places. I've proposed that only photons and particles are time-stamped with the Relativity of Simultaneity, because I do not want to believe it possible to change the time of an event merely by moving toward or away from it.

Einstein wrote that events originate at different times depending on the relative motion of each object toward or away from the event:

Every reference body (co-ordinate system) has its own particular time; unless we are told the reference-body to which the statement of time refers, there is no meaning in a statement of the time of an event.

People like us, who believe in physics, know that the distinction between past, present, and future is only a stubbornly persistent illusion.

Einstein developed this insight from his thought experiment known as The Train and Embankment. Its idea, shown above as taken from his book of 1920, is that two lightning bolts strike an embankment while a train is moving alongside it. Two observers, whose positions coincide at the moment of the strikes will see the flashes after they propagate. Observer "M" is stationary on the embankment, while "M-prime" is moving with the train. The lightning bolts hit at equal distances in front of and behind them. Knowing that the speed of light is constant, they can deduce the times the strikes occurred. I believe Einstein's conclusion that the time of events change according to the motion of objects receiving them is overdrawn.

I want to bring Einstein's car into the picture, so I'll place it on an auto-carrying train with Einstein sitting behind the wheel. Of course, Albert and Elsa could also be passengers riding in the train without the car. But since it's in the rest of the book, I've made a place for it here.

Looking at the diagram above, the flashes from the lightning bolts will reach "M" (Einstein Green) on the embankment simultaneously because he is stationary relative to the impact points. "M-prime" (Einstein Red) is moving toward the impact point ahead, and away from the impact point behind. Einstein Red will therefore see the photons from the lighting bolt ahead before he sees the photons from the lighting bolt behind. How can this be, if both bolts impacted the same distance ahead and behind him and light travels at the same speed for "M-prime" Red Einstein as it does for "M" Green Einstein? As we've seen, the Relativity of Simultaneity must adjust the time of the strikes for Red Einstein to account for his motion relative to the impact points.

Comprehending this thought experiment can be confusing because it is sometimes carelessly described as lightning bolts hitting the front and back of the train, not the embankment beside it, the way Einstein visualized it. It also shows a Relativistic process as one instantaneous event rather than evolving instant-by-instant. It would be like trying to make sense of a movie with all the frames displayed on the screen simultaneously. This is what makes Relativistic processes so difficult to understand, because they are presented as happening all at once, whereas they are created incrementally, moment by moment.

Also, we never see objects directly, but only the wave fronts of photons emitted from them at each instant of time. This doesn't matter in everyday life because we are moving slowly relative to the speed of light. But it does complicate thought experiments, when we are mentally slowing light down to speeds we comprehend in every day life, such as the time it takes to pass the length of a train. We're accustomed to seeing Newtonian motion and have no instinct for visualizing Relativistic motion where perceived time of origin changes. And finally, we must be careful to maintain a consistent perspective. Misapprehensions of Relativity happen when a perspective jumps back and forth between stationary and moving parties or jumps to a third-party perspective independent of either.

With those caveats in mind, let's try to visualize the Train and Embankment thought experiment as Einstein did. I'll use the same 8 miles, 8 minutes scenario as I did for the stand-alone Einstein's Car. The train will have to be 16 miles long, with Einstein's Car riding along at mile 8 in the middle of it. As with the car before, the train will be moving at 30mph, half the car-centric speed of light:

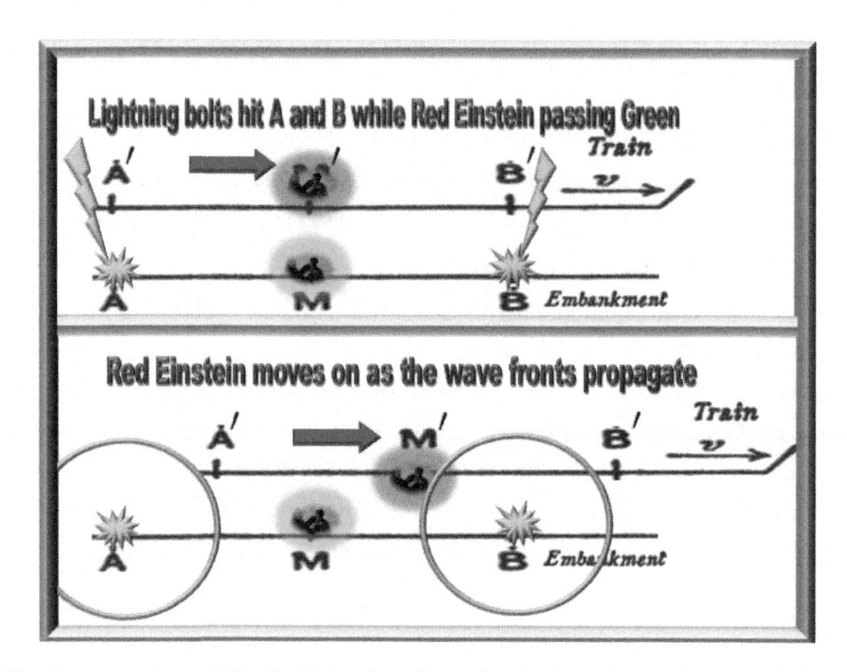

The impact points of the bolts hitting the embankment are stationary relative to Green Einstein, so each will take 8 minutes to arrive in the middle of the embankment at 12:08. Red Einstein sees the bolts as non-simultaneous:

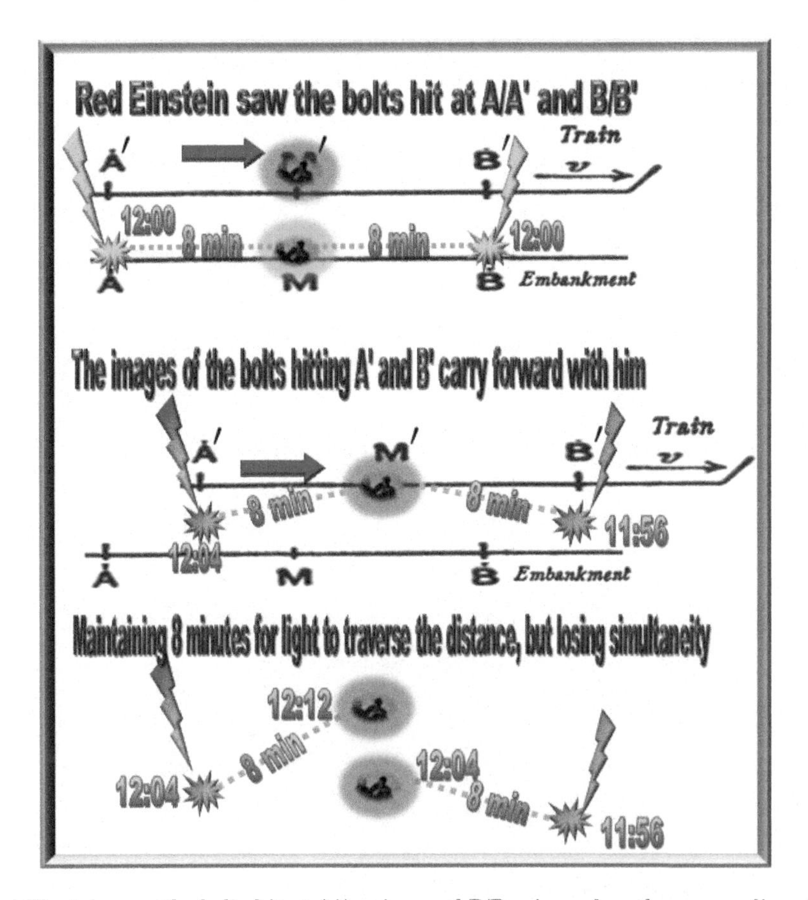

Red Einstein sees the bolts hit at A/A-prime and B/B-prime when they were aligned. Because Red Einstein is in the moving "primed" frame, he must see the impact point of the bolt at A-prime behind him track forward in time from 12:00 to 12:04 to correspond with his new position that moved four light-minutes forward. He must see the impact point of the bolt at B-prime in front of him regress from 12:00 to 11:56 to correspond with his new position of B-prime, carried forward four light minutes as he moved.

I believe there is only one time and place for each event of each lighting strike. The time of that event does not change, only the perception by Red Einstein that the time of origin of the photons changed. This is a sort of propagation-in-reverse, whereby the constant speed of light requires that the perceived time of origin of the photons must stay in synch with the relative motion of the receiving object while the photons are in flight. Events only seem to appear on different timelines because the perception of each moving object is a lookback process from each moment when photons and particles are received. The lookback forces the time of origin to *appear* to be the distance covered by the photons and particles travelling at the speed of light, from the time of reception looking backward to the time of origin. Viewed this way, the contradictions of time travel go away, because each event has one, and only one, time and place of origin.

FTL and Time Travel?

I view time travel --- in the sense of being able to change or elide events that have already appened --- as impossible due to the irreversible effects spreading out into the Universe. Even if you ad faster-than-light communications while moving at high speed, you'd only succeed in stamping he photons with a time before they happened in the timeline of stationary observers, not changing he time of the event that spawned them.

I also don't believe FTL travel or communications can be transmitted through space, because pace defines the propagation speed of *c* for electromagnetic waves of energy, force carrying particles. ecause mass is tightly bound energy, no object of mass can reach, let alone exceed, light speed. If e could travel outside space by wormholes, warp drives, higher dimensions, or by tunnelling rough space to remove the fields that permeate it, we might be able to travel, or at least mmunicate, faster than light. We might even be able to create a bubble of non-space around the ip, enabling it to move without encountering the frictions of space (which is not really all that mpty), in the way a super-cavitating torpedo can theoretically move through water at nearly the eed of sound by releasing a bubble of compressed air in front of it to keep the friction of water way from its skin. But these would not be Relativistic processes that alter time. They would be like jet travelling faster than Mach 1 getting ahead of the sound of a thunderclap without being able to ange the lighting bolt that spawned it.

A final objection to time travel is that we don't know where anything was in the past because e can't account for all the known and unknown motions of the Universe. If we could jump back in me 100 years, we would end up in some unknown point in space the Earth has moved far away om. Time is a dependency of space, but space is not a dependency of time, so it would be impossible backtrack through space, using time as a sort of autopilot to get back to the precise spatial ordinates of a time past.

I see time travel as a barren idea, even if faster-than-light travel by non-Relativistic means one day developed.

Conclusions

By reading this book, you will hopefully have gleaned a foundation for understandin Relativity and some of its unsettled controversies. If your study deepens, you should be less confuse by the headache-inducing discussions wrapped around it. And though you should be respectful c Relativity's long-established dogmas, you should seek to understand them on your terms, not jus parrot the platitudes you hear from somebody else.

I believe Relativity gets sidetracked into an overemphasis of time dilation and the fantasy c time travel, while Relativity of Simultaneity, the most fundamental consequence of Relativity constant speed of light, is under-discussed. Whenever it is disclosed to the public, it is in th grandiose terms of being a bridge between a past, present, and future allegedly coexistin simultaneously, when perhaps it is merely a time stamp on photons and force-carrying particle transmitting effects of events through space. These and other takeaways I have gleaned from m recent years of Relativity study are:

1) There is no actual or theoretical possibility of time travel to recast the past. Even cannot be undone once photons and other particles are out of their holsters and begi interacting with the Universe to propagate the change.
2) Relativity of Simultaneity does not create monolithic time-bending planes across th Universe that bring the past and future into the present as soon as you put you spaceship in gear. It is an incremental process happening instant by instant as you mov through space, backtracking and forward-tracking only the photons in flight while yo are moving toward or away from the emitting objects.
3) There is no block-time Universe where the future already exists in a predetermined wa starting 14 billion years ago at the time of the Big Bang. The future evolves from th present in deterministic and non-deterministic ways, some predictable, and som random. Our conscious minds are non-computable, so we change our environs in way physics cannot predict.
4) Time is the physical process of photons and particles of force migrating away fro events, not a mystical 4th Dimension that can be manipulated by humans, although tim does have the mathematical character of being a dimension when referencing (x,y,z, coordinates in the Universe.
5) The Universe evolves through time, and time evolves through the Universe, as carried b photons and force-carrying particles. The Universe is the body of matter, and time is i circulatory system.

6) Time dilation is the process of slowing your time by leaving a light trail across space.
7) The Universe is Relativistic in the microcosm, but possibly Machian in the macrocosm. If the Universe consisted of only two particles, we could not define a state of motion between them, as to which would be stationary in reference to the other. But we could define their state of absolute motion relative to the red / blue shift of light frequencies, and in relation to the CMB dipole.
8) When Relativity is properly understood in these terms, the so-called paradoxes of Relativity, such as The Andromeda Paradox and the Twins Paradox, resolve.

Bonus Chapter: The Humanity of Albert Einstein

What is it about Albert Einstein that makes him so widely regarded as the preeminent scientific mind of the 20th Century? Of course, there is his profound scientific insight whereby he devised the mass / energy equivalence formula $e = mc^2$ leading to his letter 30 years later to President Franklin Roosevelt on the feasibility of building the atomic bombs that concluded WWII. Combined with his collecting the bits and pieces of pre-Relativity theory going back to the 1880s when the constant speed of light became known, then organizing those bits and pieces into the well-structured Special and General Theories of Relativity. Combined with his other innovations such as probability analysis of physics phenomena and interpretation of some aspects of quantum mechanics. Einstein placed his umbrella over the complete spectrum of physics from the subatomic world to the Cosmos, during that unique window of discovery when we comprehended the inner workings of the Universe.

But there was more to him than scientific genius. He is remembered as an affable, light-hearted, humble man, without a trace of pomposity. He immersed himself in the real world of life and love as well as the scientific one. On a visit to the Grand Canyon, he was photographed dressed as an Indian and smiling with the chief to celebrate the culture of the original inhabitants. He came of age in the Pre-WWI European world of elegance and culture, soon to be annihilated by wars of aggression and genocide, ended by the atomic bombs whose theory he pioneered. He enjoyed his middle and elderly years as a free-wheeling American citizen Wherever he went, he seemed at home.

His personal life was filled with the typical triumphs and tribulations we all face. His difficult first marriage ended in divorce when he became romantically involved with his cousin Elsa. Who later passed away, leaving him a widower in his last 19 years. His first child, a daughter, was born before his first marriage, and is believed to have died of scarlet fever in a distant town before he ever saw her. Of his two sons, one became schizophrenic and spent much of his life in a mental institution in Switzerland. His other son Hans Albert followed him to America and collected soil samples for the U.S. Department of Agriculture at Clemson University in South Carolina, then became a renowned civil engineer at the California Institute of Technology and the University of California, Berkley, specializing in sediment transport, the study of how mud and sand is stirred up from river bottoms when bridges and dams are built.

In 1936, Hans Albert obtained his PhD degree at the ETH [Switzerland's national technology academy]. From 1947 to 1971 he was professor of hydraulic engineering at the University of California in Berkeley. About his father's influence on him, he once remarked, 'Probably the only project he ever gave up on was me. He tried to give me advice, but he soon discovered that I was too stubborn and that he was just wasting his time.'

Pais, Abraham. Subtle is the Lord (p. 453). OUP Oxford. Kindle Edition.

Albert is said to have denigrated his son's sediment-settling career (although it was of utmost importance to the great civil engineering projects of dams and bridges) and to have intensely disliked his daughter-in-law. Albert and Hans Albert reconciled later in life, as fathers and sons often do. The elder Einstein spent his last years among his family and friends, including his younger sister, who followed him to the United States and lived her last years with him. He advocated in vain for supranational government to end the menace of nuclear war and declined a request by Israel's government to become its president. His mind remained active through his last year, when he learned he was destined to die from an aneurism in his stomach. He rejected any artificial means of life support.

'I want to go when I want. It is tasteless to prolong life artificially; I have done my share; it is time to go. I will do it elegantly.'

Pais, Abraham. Subtle is the Lord (p. 477). OUP Oxford. Kindle Edition.

If Albert Einstein is regarded as the preeminent scientific mind of the 20th Century, perhaps it is because he was one of its preeminent people.

Further Study

Much of my work on this book has been devoted to reducing thousands of sources in books, on-line physics forums, and Youtube videos into 80 pages of coherent, easily comprehended information essential to understanding Relativity without cluttering it up --- a task even Einstein himself did not accomplish well, in my view, in his popular book of 1920. His mind was so far above ours that what he thought of as simplification hovers just beyond our mental grasp. As well, he did not have the interactive graphics tools we have to animate his ideas, so he had to make do with crudely drawn sketches.

Having sought to eliminate the clutter obscuring Relativity, I will not now clutter it by listing all the myriad sources in print and on-line that went into writing it. As for books, I have not found the recent popular ones useful, as they frequently obscure Relativity with dogma that I don't believe the authors explain well. I will therefore recommend only three books I found unique and interesting:

Subtle is the Lord (one of Einstein's noteworthy expressions) by Abraham Pais combines the fascinating story of Einstein's personal life with the physics of Relativity. Pais was Einstein's close friend and confidante. He writes the book as if you were standing in Einstein's presence. Pais was a physicist, and the math he imbeds in the book is daunting. You can skip the mathematics and just study Einstein the man, working in the most monumental time of physics and cosmological discovery we have ever known.

https://www.amazon.com/Subtle-Lord-Science-Albert-Einstein-ebook/dp/B00JI2IF62/

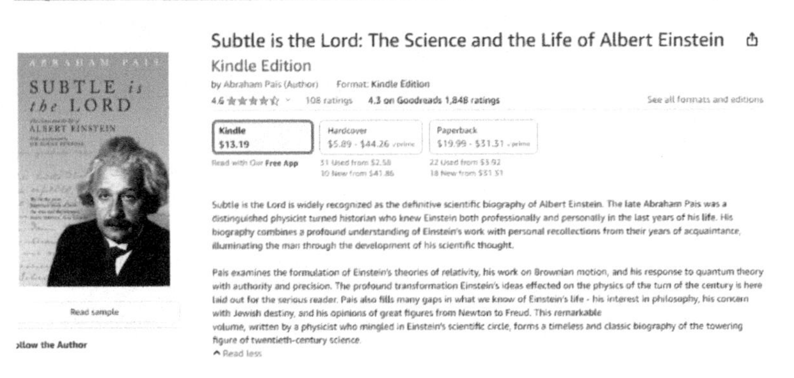

I also like the obscure book ***Relativity and Simultaneity*** by Ilya Kogan, for reasons beyond the science:

https://www.amazon.com/RELATIVITY-SIMULTANEITY-explanation-Ilya-Kogan-
book/dp/B07RHG239Z/

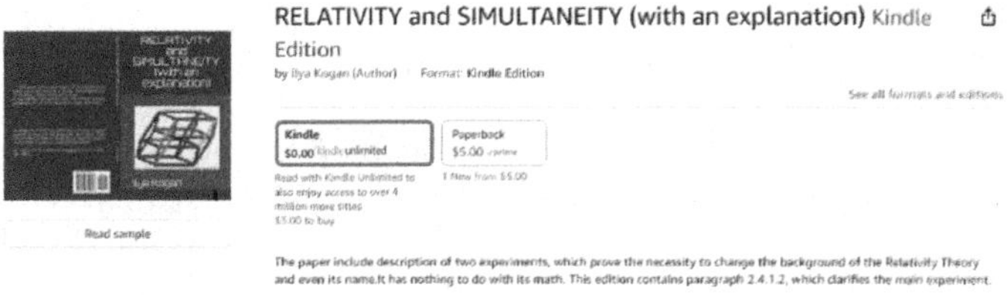

RELATIVITY and SIMULTANEITY (with an explanation) Kindle Edition

by Ilya Kogan (Author) Format: Kindle Edition

See all formats and editions

Kindle $0.00 Kindle unlimited

Paperback $5.00

Read with Kindle Unlimited to also enjoy access to over 4 million more titles $5.00 to buy

1 New from $5.00

Read sample

The paper include description of two experiments, which prove the necessity to change the background of the Relativity Theory and even its name.It has nothing to do with its math. This edition contains paragraph 2.4.1.2, which clarifies the main experiment.

The book aligns with my view of Einstein:

"Einstein once said that unless a theory can be explained to a child, the theory was probably seless; that is, the essence of a theory has to be captured by a physical picture. So many physicists get ›st in a thicket of mathematics that leads nowhere. However, like Newton before him, Einstein was bsessed by the physical picture; the mathematics would come later.

Kogan, Ilya. RELATIVITY and SIMULTANEITY (with an explanation) (pp. 6-7). Kindle dition.

The gem of the book is in its autobiographical section of Kogan LIFE (APPENDIX). It is a onderful, and sometimes tragic, story of what a young Ukrainian learned about Relativity in the 930s; then had his life upended by the German invasion of WWII, that shattered his family; then orked a career in physics and electronics in the Soviet Union; then emigrated to New York City in 986 and lived as an American through 2018, the year this book is published. Though obscure, ogan's life mirrors Einstein's as a physicist who lived at the center of the momentous events of the)th Century in Europe and the United States, including poignant life memories such as:

Thinking about my life I remember my mom, she came from the (Soviet Union) military office 945), where to her was presented a paper about the death of David (my middle brother David). uddenly, I heard a terrible groan, cry, or howl. She sang, "In vain the old lady is waiting for her son ıck home, she wept ..." Yes, at 50 she became an old, old, woman. Very old, very old, she hunched, ıd her face became wan and drawn in a few minutes.

For every mother, her son endlessly dear. The Commander can send millions into the fire Rzhev) or into the waters of Volga; maybe one out of a hundred will reach Stalingrad. Then he ılmly says as [Soviet General] Zhukov did, "Women give birth to new."

Kogan, Ilya. RELATIVITY and SIMULTANEITY (with an explanation) (p. 32). Kindle dition.

It is the life of a physicist who experienced the spectrum of good and evil, oppression an opportunity, in the 20th and early 21st Centuries, along the same lines as Einstein.

Finally, I will recommend a skeptical book on Relativity written by an author whos reputation I do not know, but who has a command of physics, and who challenges the dogma o Relativity. I do not like books seeking to debunk Relativity, as they are mostly wrapped aroun ignorance. This book takes a properly skeptical but respectful view of some of Relativity's axioms les settled than we've been led to believe. The math is formidable, but may be skipped, because th essence of the book is its questioning of Relativity's staple "thought experiments" like The Twin Paradox and The Train and Embankment, which are not as definite as they are made out to be i Relativity's dogma.

The Mechanics of Lorentz Transformations, Taha Sochi:

https://www.amazon.com/Mechanics-Lorentz-Transformations-Taha-Sochi/dp/1727118480/

As for Youtube Videos, out of the dozen or so popular Relativity theorists, I only found one - the Science Asylum videos --- worthy of recommending to a lay person. Though zany, ***Scienc Asylum*** produces well-grounded, scientifically sound videos more insightful into Relativity than th myriad others I have watched lo these many years. My favorite is his analysis of the Twins Parado the way I had previously interpreted it:

https://www.youtube.com/watch?v=UInlBJ4UnoQ

There is also the delightfully obscure **Uncanny Relativity** series by "Uncertainty drive" that open a window into Relativity's Spacetime glories:

https://www.youtube.com/watch?v=byIfbzcyjAc

https://www.youtube.com/watch?v=-eqBE1ksAEs

https://www.youtube.com/watch?v=gVAEPNLrWHY

https://www.youtube.com/watch?v=AjJwnsc4D40

https://www.youtube.com/watch?v=at7xIQ7c1yQ

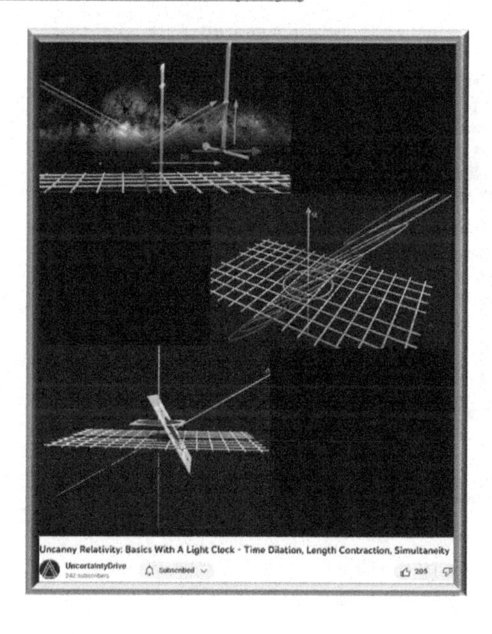

I also like the virtual light-speed journey across the Universe by "ScienceClic English" describing the odd phenomenon encountered as one approaches light speed:

https://www.youtube.com/watch?v=vFNgd3pitAI

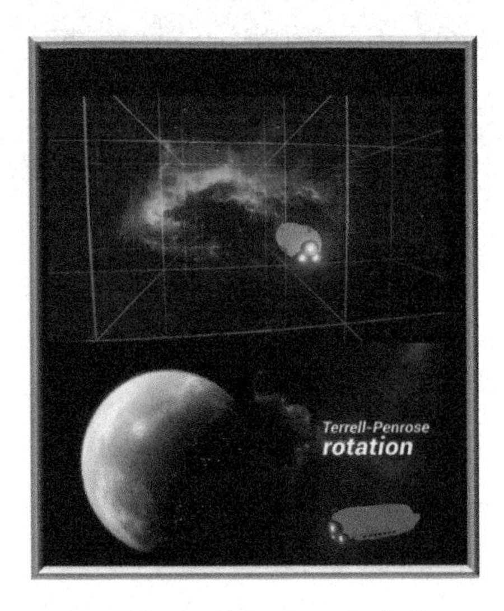

Some on-line links, starting with the links for calculating travel times to interstellar and intergalactic destinations near and far:

https://www.omnicalculator.com/physics/space-travel

https://spacetravel.simhub.online/

Space travel calculator

Home | Calculator | Difficulties of space travel | Hollywood

Do you want to travel to another planet? Or perhaps even another star system?

Then you can use this calculator to work out how long it will take you, how much energy your spacecraft needs and what your maximum velocity will be. If you travel close to the speed of light, you can also see how much time it will take from your point of view and from the point of view of the people on earth. You can also see how the length of your spacecraft will shorten for observers watching it from earth, if only they had powerful enough telescopes.

This is the simplest way to use the space travel calculator:

1. Enter a distance to a planet or star. Don't know any? Then type *Pr* and press the down arrow. The distance to Proxima Centauri appears. Select it and the distance will be filled in. Try other places in space.
2. Click ***Calculate***. The calculator determines the remaining unfilled values.
3. Click *Run*. Watch the space rocket travel from earth to your destination. Also watch the clocks of the observer and the traveler.

And the most practical program for calculating the Lorentz Transformation for Relativistic speeds:

http://www.trell.org/div/minkowski.html

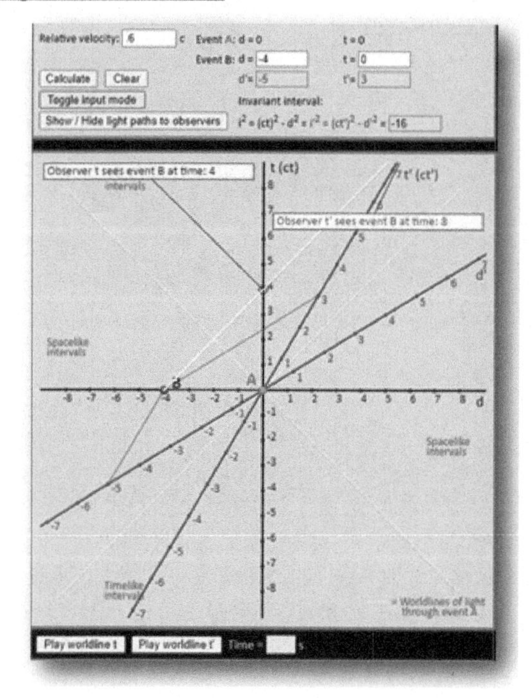

I will conclude the book, where I began it, with Einstein's popular work on Relativity, now in its 104th year of English language publication that spawned many discussions of Relativity echoing through time to the present day.

https://www.ibiblio.org/ebooks/Einstein/Einstein_Relativity.pdf

RELATIVITY

THE SPECIAL AND GENERAL THEORY

ALBERT EINSTEIN, Ph.D.

ROBERT W. LAWSON, M.Sc.

NEW YORK
HENRY HOLT AND COMPANY

Feedback: alsnewideas@gmail.com

Made in the USA
Coppell, TX
16 August 2023